Supporting Air and Space
Expeditionary Forces

Lessons from

Operation
Enduring
Freedom

Robert S. Tripp, Kristin F. Lynch
John G. Drew, Edward W. Chan

Prepared for the
United States Air Force

RAND
Project AIR FORCE

The research reported here was sponsored by the United States Air Force under Contract F49642-01-C-0003. Further information may be obtained from the Strategic Planning Division, Directorate of Plans, Hq USAF.

Library of Congress Cataloging-in-Publication Data

Supporting air and space expeditionary forces : lessons from Operation Enduring
Freedom / Robert S. Tripp ... [et al.].
   p. cm.
   "MR-1819."
   Includes bibliographical references.
   ISBN 0-8330-3517-7 (pbk. : alk. paper)
   1. United States. Air Force—Supplies and stores. 2. Airlift, Military—United
States. 3. Operation Enduring Freedom. 4. Afghanistan—History—2001– 5.
Operation Allied Force, 1999. 6. Kosovo (Serbia)—History—Civil War, 1998–1999.
7. Deployment (Strategy) 8. Logistics. I. Tripp, Robert S., 1944–.

UG1103 .S867 2004
958.104'6—dc22

                                                                    2003024740

Cover photo: Associated Press Photo at
http://www.boeing.com/news/frontiers/archive/2002.

RAND is a nonprofit institution that helps improve policy and decisionmaking through research and analysis. **RAND®** is a registered trademark. **RAND**'s publications do not necessarily reflect the opinions or policies of its research sponsors.

Cover design by Stephen Bloodsworth

Published 2004 by **RAND**
1700 Main Street, P.O. Box 2138, Santa Monica, CA 90407-2138
1200 South Hayes Street, Arlington, VA 22202-5050
201 North Craig Street, Suite 202, Pittsburgh, PA 15213-1516
RAND URL: http://www.rand.org/
To order **RAND** documents or to obtain additional information,
contact Distribution Services: Telephone: (310) 451-7002;
Fax: (310) 451-6915; Email: order@rand.org

Since 1997, the RAND Corporation has studied options for configuring a future Agile Combat Support (ACS) system that would enable the goals of the Air and Space Expeditionary Force (AEF) to be achieved. Operation Enduring Freedom (OEF), in Afghanistan, offered an opportunity to examine the implementation of new ACS concepts in a contingency environment. In 2000, RAND Project AIR FORCE helped evaluate combat support lessons from Joint Task Force Noble Anvil (JTF NA),[1] the U.S. component of Operation Allied Force (OAF), in Serbia. Some of the concepts and lessons learned from JTF NA were implemented in supporting OEF.

Supporting Air and Space Expeditionary Forces: Lessons from Operation Enduring Freedom presents an analysis of combat support experiences associated with Operation Enduring Freedom and compares these experiences with those associated with Operation Allied Force. The analysis presented an opportunity to compare findings and implications from JTF NA and OEF. Specifically, the objectives of the analysis were to indicate the performance of combat support in OEF, examine how ACS concepts were implemented in OEF, and compare JTF NA and OEF experiences to determine similarities and applicability of lessons across experiences and to determine whether some experiences are unique to particular scenarios.

---

[1] Joint Task Force Noble Anvil was the organization overseeing U.S. forces involved in Operation Allied Force. This report concentrates on Air Force operations conducted by Joint Task Force Noble Anvil.

This analysis concentrates on U.S. Air Force operations in support of OAF—specifically, Joint Task Force Noble Anvil and the first 100 days of OEF. The report focuses on experiences from OEF and what those experiences imply for a combat support system designed to ensure that AEF goals can be achieved. It does not address other portions of the War on Terrorism, such as homeland defense (for example, Operation Noble Eagle).

Task Force Enduring Look (AF/CVAX) sponsored this research, which was conducted in the Resource Management Program of RAND Project AIR FORCE, in coordination with the Air Force Deputy Chief of Staff for Installations and Logistics (AF/IL) and the Air Force Deputy Chief of Staff for Air and Space Operations (AF/XO). The research for this report was completed in February 2003.

This report should be of interest to logisticians, operators, and mobility planners throughout the Department of Defense, especially those in the Air Force.

This study is one of a series of RAND reports that address ACS issues in implementing the AEF. Other publications in the series include the following:

- *Supporting Expeditionary Aerospace Forces: An Integrated Strategic Agile Combat Support Planning Framework*, Robert S. Tripp, Lionel A. Galway, Paul S. Killingsworth, Eric Peltz, Timothy L. Ramey, and John G. Drew (MR-1056-AF). This report describes an integrated combat support planning framework that may be used to evaluate support options on a continuing basis, particularly as technology, force structure, and threats change.

- *Supporting Expeditionary Aerospace Forces: New Agile Combat Support Postures*, Lionel Galway, Robert S. Tripp, Timothy L. Ramey, and John G. Drew (MR-1075-AF). This report describes how alternative resourcing of forward operating locations (FOLs) can support employment timelines for future AEF operations. It finds that rapid employment for combat requires some prepositioning of resources at FOLs.

- *Supporting Expeditionary Aerospace Forces: An Analysis of F-15 Avionics Options*, Eric Peltz, H. L. Shulman, Robert S. Tripp, Timothy L. Ramey, Randy King, and John G. Drew (MR-1174-AF).

This report examines alternatives for meeting F-15 avionics maintenance requirements across a range of likely scenarios. The authors evaluate investments for new F-15 avionics intermediate shop test equipment against several support options, including deploying maintenance capabilities with units, performing maintenance at forward support locations (FSLs), or performing all maintenance at the home station for deploying units.

- *Supporting Expeditionary Aerospace Forces: A Concept for Evolving to the Agile Combat Support/Mobility System of the Future*, Robert S. Tripp, Lionel A. Galway, Timothy L. Ramey, Mahyar A. Amouzegar, and Eric Peltz (MR-1179-AF). This report describes the vision for the ACS system of the future based on individual commodity study results.

- *Supporting Expeditionary Aerospace Forces: Expanded Analysis of LANTIRN Options*, Amatzia Feinberg, H. L. Shulman, L. W. Miller, and Robert S. Tripp (MR-1225-AF). This report examines alternatives for meeting Low Altitude Navigation and Targeting Infrared for Night (LANTIRN) support requirements for AEF operations. The authors evaluate investments for new LANTIRN test equipment against several support options, including deploying maintenance capabilities with units, performing maintenance at FSLs, or performing all maintenance at continental United States support hubs for deploying units.

- *Supporting Expeditionary Aerospace Forces: Alternatives for Jet Engine Intermediate Maintenance*, Mahyar A. Amouzegar, Lionel A. Galway, and Amanda Geller (MR-1431-AF). This report evaluates the manner in which Jet Engine Intermediate Maintenance (JEIM) shops can best be configured to facilitate overseas deployments. The authors examine a number of JEIM support options, which are distinguished primarily by the degree to which JEIM support is centralized or decentralized.

- *Reconfiguring Footprint to Speed Expeditionary Aerospace Forces Deployment*, Lionel Galway, Mahyar A. Amouzegar, R. J. Hillestad, and Don Snyder (MR-1625-AF). This study develops an analysis framework—footprint configuration—to assist in evaluating the feasibility of reducing the size of equipment or time-phasing the deployment of support and relocating some equipment to places other than forward operating locations. It also attempts to define *footprint* and to establish a way to monitor its reduction.

- *Supporting Expeditionary Aerospace Forces: An Operational Architecture for Combat Support Execution Planning and Control*, James A. Leftwich, Robert S. Tripp, Amanda Geller, Patrick H. Mills, Tom LaTourrette, Charles Robert Roll, Cauley Von Hoffman, and David Johansen (MR-1536-AF). This report outlines the framework for evaluating options for combat support execution planning and control. The analysis describes the combat support command and control operational architecture as it is now and as it should be in the future. It also describes the changes that must take place to achieve that future state.

RAND PROJECT AIR FORCE

RAND Project AIR FORCE (PAF), a division of the RAND Corporation, is the U.S. Air Force's federally funded research and development center for studies and analyses. PAF provides the Air Force with independent analyses of policy alternatives affecting the development, employment, combat readiness, and support of current and future aerospace forces. Research is conducted in four programs: Aerospace Force Development; Manpower, Personnel, and Training; Resource Management; and Strategy and Doctrine.

Additional information about PAF is available on our web site at http://www.rand.org/paf.

CONTENTS

TABLES

The Air and Space Expeditionary Force (AEF) concept was developed
to enable the Air Force to respond quickly to any national security issue with a tailored, sustainable force. The major theme of substituting speed of deployment and employment for presence has significant resource implications. Since 1997, the RAND Corporation and
the Air Force Logistics Management Agency have studied and refined
a framework for an Agile Combat Support (ACS) system to support
the AEF concept (Galway et al., 2000; Tripp et al., 1999).

AGILE COMBAT SUPPORT SYSTEM BACKGROUND

As described in Tripp et al. (2000), the AEF operational goals are to

- rapidly configure support needed to achieve the desired operational effects

- quickly deploy both large and small tailored force packages with
  the capability to deliver substantial firepower anywhere in the
  world

- immediately employ such forces upon arrival

- smoothly shift from deployment to operational sustainment

- meet the demands of small-scale contingencies and peacekeeping commitments while maintaining readiness for potential contingencies outlined in defense guidance.

Key elements of an ACS system to enable these AEF operational goals
include the following (Tripp et al., 1999):

- A combat support execution planning and control (CSC2) system to assess, organize, and direct combat support[1] activities, meet operational requirements, and be responsive to rapidly changing circumstances. The CSC2 capability would help combat support personnel

  — Estimate combat support resource requirements and process performances needed to achieve the desired operational effects for the specific scenario.

  — Configure supply chains for deployment and sustainment, including the military and commercial transportation needed to meet deployment and sustainment needs.

  — Establish control parameters for the performance of various combat support processes required to meet specific operational needs.

  — Track actual combat support performance against control parameters.

  — Signal when a process is outside accepted control parameters so that plans can be developed to get the process back within control limits.

- A quickly configured and responsive distribution network to connect forward operating locations (FOLs), forward support locations (FSLs), and continental United States (CONUS) support locations (CSLs)

- A network of FOLs resourced to support varying deployment/employment timelines

- A network of FSLs configured outside CONUS to provide storage capabilities for heavy war reserve materiel (WRM), such as munitions and tents, and selected maintenance capabilities, such as centralized intermediate repair facilities (CIRFs) that service jet engines of units deployed to FOLs. FSLs could be collocated with transportation hubs.

---

[1]In this report, the term *combat support* is defined as anything other than the actual flying operation. Combat support consists of civil engineering, communications, security forces, maintenance, service, munitions, etc.

- A network of CSLs, including Air Force depots, CIRFs, and contractor support facilities. As with FSLs, a variety of different activities may be set up at major Air Force bases, convenient civilian transportation hubs, or Air Force or other defense repair depots.

In 2000, Project AIR FORCE helped evaluate agile combat support lessons from Joint Task Force Noble Anvil (JTF NA),[2] in Serbia. Some of the concepts and lessons learned from JTF NA were implemented in supporting Operation Enduring Freedom (OEF), in Afghanistan. This analysis allowed the opportunity to compare findings and implications from JTF NA and OEF. Specifically, the objectives of the analysis were to indicate how combat support performed in the OEF scenario, examine how ACS concepts were implemented in OEF, and compare JTF NA and OEF experiences to determine similarities and applicability of lessons across experiences and to determine whether some experiences are unique to particular scenarios.

JTF NA and OEF provide important opportunities to study how AEF ACS concepts were implemented during contingency operations and how they have been refined to better support AEF goals. In this report, we address five areas that correspond to the above elements of an ACS system: CSC2 structure, the development of forward operating locations, the use of forward support locations and CONUS support locations, the transportation system, and resourcing to meet current operational requirements. Understanding these experiences could be of value for combat support and operational personnel who may be called upon to support future contingency operations. Task Force Enduring Look (AF/CVAX) sponsored this research in coordination with the Air Force Deputy Chief of Staff for Installations and Logistics (AF/IL).

COMBAT SUPPORT CHARACTERISTICS OF JTF NA AND OEF

Since every military operation has its own unique characteristics, neither the performance of the current support system nor the de-

---

[2]Joint Task Force Noble Anvil was the organization overseeing U.S. forces involved in Operation Allied Force. This report concentrates on Air Force operations conducted by Joint Task Force Noble Anvil.

sign of a future combat support system should be assessed or based solely on any one experience. However, both JTF NA and OEF provide important experiences that warrant study for combat support operations.

By some measures, OEF could be considered a small combat operation, given the number of aircraft and personnel deployed, the number of beddown locations employed,[3] and the number of sorties flown, all of which are small compared with other recent Air Force operations (see Table S.1). However, the combination of short planning timelines and poor existing infrastructure created especially demanding requirements for combat support operations. By comparison, JTF NA had the benefit of a long buildup time and was conducted from bases with good infrastructure.

Since, as the Chief of Staff of the Air Force recently observed, "[The Air Force's] heightened tempo of operations is likely to continue at its current pace for the foreseeable future" (Jumper, 2002), the Air Force must be able to support the deployment of a large number of forces, either in large-scale deployments, such as in Operation Desert Storm, or in an accumulation of a number of small-scale contingencies, such as JTF NA and OEF. Furthermore, it must be able to provide such support on short notice and in austere environments,

Table S.1

Dimensions of Support in JTF NA and OEF

	JTF NA	OEF
Number of Air Force aircraft deployed	500	200
Number of Air Force personnel deployed	44,000	12,000
Number of munitions expended (in short tons)	7,000	7,000
Number of beddown locations	25	14
Number of sorties flown	30,000	11,000

---

[3]We use the term *beddown location* to refer to locations at which personnel and/or aircraft were based during operations.

particularly as the War on Terrorism continues around the world. For these reasons, JTF NA and OEF are worthy of study.

In both JTF NA and OEF, the effectiveness of combat support was due, in part, to many ad hoc innovations and adaptations made by combat support personnel to overcome shortcomings in doctrine, training, organizations, and systems, and to shortfalls in resources. These shortcomings and shortfalls can be placed into four general categories: CSC2 doctrine, training, and tools; forward operating location development; the role of the Air Force in joint activities, including development of the theater distribution system (TDS) and installation construction; and the misalignment of resource-planning assumptions and the realities of resource employment of today's boiling peace and contingency operations. We discuss the findings and implications for each of these categories in turn, below.

COMBAT SUPPORT EXECUTION PLANNING AND CONTROL

Findings (see pp. 19–32)

Processes for combat support execution planning and control and organizational alignments have improved since JTF NA, but OEF shows that more attention is still needed in this area. CSC2 was not well understood, so an ad hoc organizational structure developed that varied from doctrine and continued to evolve throughout the operation.

Lessons learned from operations in Serbia indicated problems in CSC2. As a result, AF/IL asked RAND Project AIR FORCE to develop a CSC2 future, or "TO-BE," operational architecture to address the needs of the AEF. Work began in 2000 and was concluding when operations began in Afghanistan in September 2001. Although the TO-BE operational architecture was not used during OEF, OEF provided an opportunity to test its design.

The organizational command structure of OEF combat support differed from the structure delineated in doctrine. As roles and responsibilities developed on an ad hoc basis, this difference led to several difficulties, and some organizations were not prepared for the evolving responsibilities. The global nature of OEF and other ongoing operations further complicated the command structure.

Responsibilities were distributed across commands and regions, increasing information-sharing burdens.

The necessity to prioritize among competing demands for time and resources increased with multiple commitments to ongoing operations, such as homeland defense (Operation Noble Eagle, a Northern Command [formerly Air Combat Command] responsibility); Central Command's exercise Bright Star; support of Bosnia, a U.S. Air Forces, Europe (USAFE) responsibility; Operation Northern Watch, a USAFE responsibility; Operation Southern Watch, a U.S. Air Forces, Central Command (CENTAF) responsibility; and others. Even though developed on an ad hoc basis, the command relationships developed during OEF evolved to closely resemble the CSC2 TO-BE architecture (Leftwich et al., 2002) (described in Appendix A).

Measuring the performance of combat support was another issue faced in OEF; such measurement was continuously deficient. Commercial carriers, for example, had end-to-end visibility of their assets and could track delivery time; the military did not/could not. More performance metrics were defined and tracked in OEF than in prior conflicts, and OEF has shown an improvement in the use of combat support feedback mechanisms. However, many of these metrics did not have goals or standards established to gauge performance levels of combat support processes against those needed to meet operational needs or requirements.

Implications (see pp. 33–34)

As a result of JTF NA experiences, for which CSC2 doctrine and training were underdeveloped, the Air Force has developed a TO-BE CSC2 operational architecture that outlines how CSC2 doctrine, training, organizations, and tools need to be developed to meet agile combat support needs. At the time of OEF, this work was just being completed and the recommendations in these areas were not fully understood or endorsed by the Air Force's senior leaders. Consequently, the Air Force developed ad hoc CSC2 command lines and organizational alignments during OEF, as it did in JTF NA.

OEF has shown an improvement in the use of combat support feedback mechanisms. Although tracking metrics represents an improvement in command and control, and better information has become available, further efforts are needed to establish a closed-loop[4] combat support control system, as described in the CSC2 TO-BE operational architecture and Appendix B.

FORWARD OPERATING LOCATIONS AND SITE PREPARATION

Findings (see pp. 35–46)

Austere FOLs and an immature theater infrastructure during OEF emphasized the importance of early planning, knowledge of the theater, and FOL preparation. Even with a more-developed infrastructure, FOL developed during JTF NA was delayed by host-nation support and site surveys. Host-nation support was difficult to negotiate. Site surveys were ad hoc and nonstandardized in both JTF NA and OEF. Resultant deployment timelines varied widely in both operations.

Timelines for force package deployment varied and depended heavily on preparation activities at forward operating locations. FOLs that were partially developed or at which the Air Force had experience from previous deployments facilitated rapid force deployments—for example, in as few as 17 days during OEF. However, FOLs selected in unanticipated locations took up to 70+ days to become fully developed. Preparation of FOLs was slowed by, among other things, establishing host-nation support, the time used to conduct site surveys, the quality of the data received from the surveys, the amount of development needed to bed down forces, and the amount of contract support available. Even when country clearance had been granted, access to specific bases was often not granted at the same time. These same issues were faced during JTF NA.

Risks were taken by the Air Force to satisfy operational objectives. During OEF, some forces were bedded down at locations that were not yet fully developed. Living conditions were sometimes unsani-

---

[4]A *closed-loop process* takes the output and uses it as an input for the next iteration of the process.

tary. To meet operational requirements, these conditions were accepted and some creature comforts were sacrificed.

During both operations, no standard site-survey tool was used. During OEF, Air Combat Command (ACC) personnel used GEOReach[5] to conduct surveys, whereas Air Mobility Command (AMC) personnel did not use an automated tool. Often, the data gathered were not shared between commands. The lack of standards and tools between the Air Force, coalition partners, and other services led to delays in developing FOLs.

Preparation efforts conducted by Civil Engineer (CE) personnel played a large role in getting OEF FOLs ready for deploying forces. Civil Engineer personnel and resources were overextended in construction efforts for ongoing Operation Southern Watch deployments, as well as in new-construction efforts to support OEF deployments. Development of installations in support of OEF was the largest such effort since Vietnam.

In addition, during OEF, the Air Force accepted more responsibility for site developments than originally planned, many times getting its assets in place more quickly than the Army did. As a result, the Air Force developed some joint sites or portions of jointly occupied sites that were originally planned for Army development. The Air Force developed more than 75 percent of the FOLs in OEF, many of which supported joint operations.

Finally, contractor support facilitated FOL development. WRM contractors, who were in place prior to OEF at forward support locations that also served as forward operating locations, were able to support FOL-development actions for some of the first units to arrive in the area of responsibility (AOR). The CENTAF WRM support contractor in the AOR assisted in initial activities associated with FOL development, including building tent cities, setting up fuel farms, operating power plants, and providing food and services for the airmen. Contractor support was vital to fast FOL development during OEF.

---

[5]GEOReach is a program that combines tabular data with a visual image to provide commanders with situational awareness.

Implications (see pp. 46–48)

Both JTF NA and OEF illustrate that more attention should be focused on political agreements and engagement policies required to develop FOLs. Successes include development of the bomber islands, such as Diego Garcia, which was used effectively in OEF. Other examples include evolving and enhancing capabilities such as those at Shaikh Isa Air Base, Al Udeid Air Base, and Al Dhafra Air Base. The Air Force has also recognized the need to develop site survey tools, standardize the procedures for collecting data on FOLs, and develop assessments of rapid-beddown capability. Some funded developments in this area—programs such as Survey Tool for Employment Planning (STEP) and Beddown Capability Assessment Tool (BCAT)—did not meet with as much success during OEF partly because these tools required classified networks and partly because of the phasing of their development cycles. Other tools, such as GEOReach, were used effectively during OEF. More can be done to standardize site-survey procedures and processes within the Air Force, with U.S. allies, and with other services. The Air Force recognizes this and has taken steps to improve these areas.

During OEF, the Air Force accepted some additional responsibilities for developing base infrastructure when the Air Force and the Army were located at sites where the Army was the responsible agent, partly because of the Air Force's "need for speed" and partly because the development assets for Army installations were palletized, preventing Army assets from being transported fast enough to build infrastructure to support Air Force operations. Executive Agency responsibility is an area that needs to be addressed by Air Force and Army combat support planners.

FORWARD SUPPORT LOCATION/CONUS SUPPORT LOCATION PREPARATION FOR MEETING UNCERTAIN FOL REQUIREMENTS

Findings (see pp. 49–56)

The current AEF force structure of light, lean, and lethal response forces is highly dependent upon FSL capacities and throughput.

Austere FOLs and the immature theater infrastructure illustrated the importance of using FSLs efficiently during OEF. Because of problems identified during JTF NA, improvements have been made in linking FSLs and CSLs to dynamic warfighter needs. But, much more can be done in this area.

Effective agile combat support enabled rapid force deployment, employment, and uninterrupted sustainment of operations in both JTF NA and OEF. As a result of JTF NA experiences, the Air Force has taken several actions to enhance agile combat support. Our analyses indicate that these actions directly contributed to the effectiveness of ACS in OEF. Air Force actions to enhance selected FSLs and develop a global strategy for positioning heavy non-unit resources have been steps in the right direction and have directly contributed to OEF combat support successes. They include selecting and resourcing an additional Afloat Prepositioned Ship (APS); putting munitions in containers on APS; sponsoring an additional WRM ship (a forward support location, afloat); creating formal maintenance FSLs (called CIRFs); and recognizing that improving throughput at WRM forward support locations is key to rapid deployments. These actions facilitated rapid deployment and sustainment of OEF operations, but more can be done.

As in JTF NA, moving assets from FSLs to the FOLs satisfied most FOL combat support requirements. But potential throughput constraints at some FSLs were uncovered during OEF that could adversely affect large force-package deployment timelines if not corrected.

During JTF NA, resource constraints such as backorders hindered the effectiveness of CONUS support locations by adding substantial resupply time and variability during the conflict. Although backorder rates improved, they remained high throughout Operation Allied Force.

CSLs were used more effectively during OEF. Because of JTF NA experiences, attention was given to creating better links between CSLs and the warfighters. Air Force Materiel Command's (AFMC's) Logistics Support Office and the High Impact Target (HIT) list—the most important repair parts for AFMC to monitor in the various Air

Logistics Centers—enhanced CSLs' responsiveness to warfighters during OEF.

Forward support locations for aircraft maintenance were used successfully during both JTF NA and OEF. JTF NA showed that preselection and resourcing—with personnel and equipment—of centralized support facilities can improve flexibility and reduce the assets necessary to deploy to an FOL. During OEF, CIRFs satisfied intermediate maintenance requirements for a number of reparable items for deployed fighter units, not only reducing the forward deployed equipment and personnel but also supporting forward bombers' phase maintenance. Goals were established to link warfighter needs to the CIRF maintenance process and theater distribution system performance. More attention needs to be placed on examining the direct linkages between the performance of the CSL combat support process and operational goals, such as that established at the CIRF.

Implications (see p. 57)

Both JTF NA and OEF proved the Air Force's enhancement of selected FSLs and development of a prepositioning strategy for WRM to be steps in the right direction. However, many FSL developments are still oriented toward specific AORs. A more global development strategy is needed to achieve a more integrated and global ACS posture.

Because of the experiences with backorders during JTF NA, the Air Force has recognized the need for and given attention to creating better links between the CSLs and the warfighters. Examples include AFMC's Logistics Support Office and High Impact Target list to enhance responsiveness to the warfighters from CSLs. However, much more can be done, in this area. For example, a closed-loop feedback capability could be developed to measure actual performance of CSL processes against those needed (planned) to achieve specific operational objectives (see Appendix B).

RELIABLE TRANSPORTATION TO MEET FOL NEEDS (THEATER DISTRIBUTION SYSTEM)

Findings (see pp. 59–72)

AEF operational goals are dependent upon capabilities for assured and reliable end-to-end deployment and distribution that can be configured quickly to connect the selected sets of FOLs, FSLs, and CSLs in contingency operations. Under current joint doctrine, the service with the preponderance of force may be delegated the responsibility for developing and operating the TDS. Since the Air Force may be the predominant user of TDS in early phases of future campaigns, the Air Force may be delegated the TDS responsibility. Even if this responsibility is delegated to another service, the Air Force should play an active role in determining TDS capacities and capabilities. AEF success depends on the early establishment of reliable and responsive TDS capabilities. The Air Force, as well as other services, depends on joint, global, multimodal, end-to-end transportation capabilities.

In both JTF NA and OEF, problems encountered with establishing a responsive TDS and the problems associated with integrating the strategic movements system with TDS led to gaps in an end-to-end military deployment and resupply system that were not encountered by commercial carriers. During OEF, Federal Express and other carriers had end-to-end visibility and could track their responsiveness in meeting deliveries. This same kind of capability was not established until several months after operations began in the military portion of the system.

Even with a well-developed transportation infrastructure in Western Europe during JTF NA, the configuration and performance of the theater distribution system required continuous refinement to ensure that supplies were delivered to operational units, and innovative approaches were taken and adaptations made to mitigate TDS shortfalls. OEF, by contrast, took place in an AOR with a very poor transportation infrastructure. Issues with TDS performance arose.

Regardless of the richness or poorness of the transportation infrastructure, TDS development and operation will require more attention in all future operations, because the Air Force relies on robust and reliable resupply and because most materiel needed to initiate and sustain operations is located at FSLs.

In OEF, Central Command tasked the Air Force to develop and manage TDS, including the Joint Movement Center—an Army responsibility in the past. Although the Air Force may not have expected the TDS role and may not have had personnel adequately trained to accomplish that role, the TDS role is vitally important to the accomplishment of AEF operational goals, including rapid deployment and uninterrupted sustainment.

As OEF unfolded, the theater distribution system continued to evolve. Fuel dominated movement requirements. Assets such as FOL support, munitions, and rations also accounted for a significant portion of movements. Although spares account for only a small portion of the transportation requirements, the light, lean, and lethal AEF depends upon rapid and reliable resupply. Another issue arose when materiel from one AOR had to be transported to another AOR.

To move all the assets required to sustain an operation, many modes of transportation are used. Distribution planners need to consider multiple modes and both commercial and military carriers when planning the end-to-end distribution network to support AEFs at deployed sites. During OEF, in addition to Air Force assets, commercial airlift and land transportation were contracted, and sealift was used. As in JTF NA, the transportation system was complex and involved coordination between services, among coalition partners, and between different AORs. No single supply chain dominated or consistently outperformed all other chains to each deployment location.

TDS has two components: In the first component, assets are moved from the FSLs to the FOLs. This part of the distribution system is required to move initial deployment and sustainment items to the point of need, and many of those items are stored in the AOR. The second component provides the onward movement of resources from CONUS and the movement of reparable parts to and from FSLs.

During both JTF NA and OEF, the complex intratheater movement system was not always well coordinated with the strategic (intertheater) movement system. Gaps exist between the two systems. Cargo stacked up at transshipment points, and a significant portion of the total customer wait time was accounted for by the time cargo spent in the transshipment hubs.

Implications (see pp. 72–73)

Another area that needs considerable attention is the Air Force's role in the development of the TDS. As JTF NA made evident, even in well-developed countries, TDS can be problematic

Joint publications and combatant commander concepts of operations indicate that the combatant commander can delegate the development of the TDS to one service, which is generally the predominant user of the system, and that service will develop the system with coordination from the other services. In the past in the Central Command AOR, this responsibility was given to the Army. In OEF, the Air Force became the predominant user in the early phases of the operation. Current joint doctrine places the responsibility for the development of the strategic movement system on the U.S. Transportation Command (USTRANSCOM). Thus, current doctrine splits the responsibility for developing the "end-to-end" deployment and resupply system among multiple parties.

In today's world of global warfare and having those combat support facilities located in one AOR supporting a combatant commander in another AOR, what could be considered TDS and what could be considered strategic movements are confused. For instance, is the system that moves WRM or repaired spares from the European Command AOR to the Central Command AOR a TDS or a strategic movement system? Current joint doctrine may be inappropriate for expeditionary forces that rely on fast deployment, immediate employment, and reliance on responsive resupply of lean, forward-deployed forces. In today's global War on Terrorism, it may be more appropriate for USTRANSCOM to develop capabilities for end-to-end distribution channels rather than split distribution responsibilities among USTRANSCOM and multiple combatant commanders.

Another solution may be to develop Distribution Units in each service. Similar to a Federal Express or a United Postal Service regional office, such units would be trained to fill in the gaps between the strategic and the tactical distribution systems. They could have common training, tools, and performance metrics and could seamlessly merge into the TDS gap during contingency operations.[6]

If new joint doctrine is developed, and even if the current doctrine is retained, the Air Force must be prepared to play a lead role in developing the end-to-end distribution channels. Reliance on light and lean deployments and responsive resupply of deployed units places great importance on the rapid development of contingency end-to-end deployment and distribution capabilities. Air Force Logistics Readiness Officers and enlisted personnel are logical candidates for carrying out this development, but they lack sufficient training to fulfill the responsibilities. Additional policies for training and personnel development are needed for the Air Force to meet these future distribution responsibilities.

RESOURCING TO MEET CONTINGENCY, ROTATIONAL, AND MAJOR REGIONAL CONFLICT REQUIREMENTS

Findings (see pp. 75–86)

Although JTF NA and OEF combat scenarios differed from each other in many ways, they also differed from wartime planning factors in significant ways. The usage factors associated with supporting permanent rotational commitments such as Operation Southern Watch and unanticipated contingency operations are different from those used to make programming decisions to obtain resources. The Air Force is operating today with a resource base that was created largely using previous guidance, for which resources were computed to meet the requirements of major regional conflicts (MRCs). Under this programming paradigm, non-MRC requirements—for example, those needed to meet permanent rotations, peacekeeping, and smaller contingency operations—were assumed to be subsets of the MRC resource base. The current combat support system and programmed resource base have difficulty simultaneously supporting small-scale contingencies and current rotational deployment requirements.

---

[6]For more information about this Distribution Unit concept, see Halliday and Moore (1994).

xxx   Supporting Air and Space Expeditionary Forces

Shortages in combat support assets, particularly in high-demand, low-density areas, such as combat communications, civil engineering, and force protection, stressed the AEF construct, resulting in the Air Force's borrowing against future AEFs during OEF. The current AEF scheduling rules, which allow personnel to be eligible for deployment for only 90 days in a 15-month cycle, may be an effective and efficient way to schedule and deploy aircraft and aircraft support units, but they may not be the best way for scheduling ACS. Specifically, balances must be struck between home-station support disruption and deployment commitments.

To evaluate combat support options in today's uncertain world requires a capability-based assessment method. Such a method provides insights into the capabilities that exist to meet a wide variety of scenarios with alternative levels of investments in combat support resources. For instance, how many bare bases can be opened and sustained within any given AEF cycle? What is the constraint, personnel or equipment? If personnel, which career fields? What are reasonable options for alleviating the constraints? A capabilities view of resources may be a more appropriate way than a scenario-specific view to consider resource investments today.

Finally, the AEF is a transformational construct and has many implications for what types of resources will be provided and how those resources will be provided in the future. The major theme of substituting speed of deployment and employment for presence has significant resource implications. It also has significant implications for the types of resources that need to be procured. The deployment of lean initial support packages to get quickly to the fight places emphasis on reliable transportation and CSC2.

Implications (see pp. 86–88)

The planning factors and assumptions that are used to determine resource requirements differ significantly from those that are encountered in current rotational and contingency operations. As found in many cases in JTF NA and OEF, the current resource-usage factors are more demanding than the assumptions used to fund resources. This imbalance creates the resource shortages that appear in contingency operations. In addition, the current AEF scheduling rules must be routinely violated in some key combat support areas, such as fire protection and combat communications. Alternatives for providing combat support capabilities to AEFs need to be reviewed. The current approach creates stress and limits combat power-projection capabilities.

CONCLUSIONS

Table S.2 summarizes combat support in the five areas investigated during this study.

Opportunities for improving combat support for the AEF of the future—for making it more congruent with agile combat support—are described in the following recommendations.

Combat Support Execution Planning and Control

• Establish clear doctrine for combat support execution planning and control.

• Clearly define command relationships.

Table S.2

Assessing Combat Service Support

	Operation Allied Force, JTF NA	Operation Enduring Freedom
Combat support execution planning and control	Ad hoc	Improved, but still ad hoc
Forward operating location development and site preparation	Varied	Varied
Forward support location and CONUS support location preparation and operation	Inefficiently used	Better linked to warfighter needs
Reliable transportation to meet forward operating location needs	Not prepared for responsibility	Inadequate; built on existing Operation Southern Watch system
Resourcing to meet contingency, rotational, and MRC requirements	Differed from planning factors	Differed from planning factors

- Integrate combat support planning with the operational campaign planning process.
- Develop control mechanisms.

FOL and Site Preparation

- Focus attention on political agreements and engagement policies.
- Standardize site-survey procedures and processes within the Air Force, with other services, and with U.S. allies.

FSL/CSL Preparation for Meeting Uncertain FOL Requirements

- Further develop the existing global network of FSLs and CSLs.
- Continue improvements in linking FSLs and CSLs to dynamic warfighter needs.

Reliable Transportation to Meet FOL Needs (TDS)

- Be prepared to play an active role in determining TDS capacities and capabilities
  - Identify lift requirements, including airlift, sealift, and movement by land
  - Initiate training and enhance personnel development policies to prepare for TDS responsibility
  - Work with joint commands to develop and resource plans to support the AEF with adequate TDS capabilities.
- Review joint doctrine on the transportation system
  - Consider having USTRANSCOM develop an end-to-end distribution channel
  - Consider establishing Distribtuion Units in each service to fill in TDS gaps during contingency operations

— Consider ways to improve TDS performance, including better in-transit visibility and demand-forcasting mechanisms.

Resourcing to Meet Contingency, Rotational, and MRC Requirements

- Reevaluate current processes and policies for AEF assignments and the current Program Objective Memorandum (POM) assumptions with respect to combat support resources

  — Align current employment practices with resource-planning factors.

- Enhance the capabilities-based planning and assessment methods that RAND is currently developing.

- Evaluate existing scheduling rules for combat support with respect to how that support will affect the performance of home-station and deployed combat support.

ACKNOWLEDGMENTS

Many individuals inside and outside the Air Force provided valuable assistance and support to our work. We separate our acknowledgments into those for individuals who helped our efforts in gathering data associated with Operation Enduring Freedom (OEF) and those who helped us in gathering data associated with Joint Task Force Noble Anvil (JTF NA).

ACKNOWLEDGMENTS FOR OEF

We thank Brigadier General James Hunt and Colonel Thomas Entwistle of Task Force Enduring Look (TFEL), AF/CVAX, for sponsoring this analysis. Mr. Richard Bird of TFEL was also very helpful in providing access to data. Without their support, we would not have been able to undertake this analysis.

We are also grateful to Lieutenant General Michael Zettler, Deputy Chief of Staff, Installations and Logistics (AF/IL) during Operation Enduring Freedom (OEF), for his support of this effort. General Zettler sponsored some of our earlier research associated with developing Agile Combat Support (ACS) concepts, asking that we specifically investigate operations in Afghanistan to determine whether this experience might change some of our ACS analyses and conclusions. Major General Jeff Kohler, Director of Operations Plans (AF/XOX) during OEF, was also instrumental in requesting our assistance in examining combat support lessons resulting from OEF and taking appropriate steps to include lessons in the design of the ACS system of the future.

We are also especially grateful to Brigadier General Robert Elder and Colonel Duane Jones for their assistance. General Elder is the Vice Commander for U.S. Air Forces, Central Command (CENTAF), Shaw AFB, South Carolina. Colonel Jones, Director of Logistics for CENTAF, Shaw AFB, South Carolina, was responsible for planning and executing all combat support operations for U.S. aerospace forces in the theater. Both General Elder and Colonel Jones provided free and open access to everyone under their commands during our search for lessons that might affect the air and space expeditionary force combat support system of the future. Colonel Jones not only supported numerous interviews with his staff, he also set up a visit to the area of responsibility (AOR) for our analysts to gain first-hand knowledge of conditions and support requirements.

At the Air Staff, we thank Ms. Susan O'Neal, AF/IL, Mr. Michael Aimone, AF/ILG, Colonel Connie Morrow, AF/ILGX, and their staffs for their support and critiques of this work. Each of these individuals played a key role in the Air Staff Combat Support Center during OEF; each has provided insights that have been instrumental in shaping this research.

We have enjoyed support for our research from the Air Force's major commands that were involved in operations in Afghanistan. We are especially grateful to Major General Don Wetekam, ACC/LG, and Brigadier General Pat Burns, ACC/CE, for their insights on how CENTAF A-4/7 Rear operated to support OEF. We thank Brigadier Generals Art Rooney, USAFE/LG, Maury Forsyth, USAFE/UTASC, Dave Gillett, USAFE/LG, Art Morrill, PACAF/LG, Pete Hennessey, AMC/LG, Pacific Command/J-4, and Claude Christianson, U.S. Forces, Korea/J-4, and members of their staffs for their support.

Many individuals at different commands assisted in our data-collection effort. At the European Command/J-4, we thank, Lieutenant Colonel Chiarotti. Major General Terry Gabreski, AFMC/LG, assisted in providing insights on AFMC support to OEF. At CENTAF, Major Dennis Long played a key role in helping to obtain data for our analysis. In addition, Major Long set up our visit to Southwest Asia and accompanied us on our visit to the AOR. Mr. Bob Lutz of DynCorp helped provide access to DynCorp personnel and data that helped complete our analysis. Technical Sergeant Julie Bitney, OO-ALC/ACP, and Senior Master Sergeant Eric Pickett,

AFLMA/LGM, assisted us in obtaining data on munitions. Chief Master Sergeant Pete Christofferson, 16 SUPS/LGSC, and Master Sergeant Sanders Louvierre, 2 SUPS/LGSPC, helped track down backlog data.

Others who have helped include: Colonel Dave Smith, PACAF/LGX, Colonel Carol King, 13th AF/LG, Colonel Russ Grunch, 7th AF/LG, Colonel Buck Jones, USAFE/LG, Major Maria Garcia, USAFE/LGX, Major Duane Bowen, AF/ILGX, Lieutenant Colonel Orlando Papucci, CENTAF/LGT, Major Tyrone Eason, USA, CENTAF/LGT, Lieutenant Colonel Lovato, CENTAF/LGX, Lieutenant Colonel David Johansen, AF/ILGX, many people at AMC/TACC, and numerous others who helped us gather data and form perspectives on OEF.

ACKNOWLEDGMENTS FOR OAF

We thank General John A. Handy, Commander of Air Mobility Command, for initiating the analysis of OAF when he was Deputy Chief of Staff, Installations and Logistics (AF/IL). General Handy sponsored some of our earlier research directed at evaluating Agile Combat Support (ACS) options for enabling AEF operational goals. From our analyses, we advanced several ACS concepts for the AEF. During his review of the AEF work, General Handy asked that we specifically investigate OAF to determine how this experience might change some of our analyses and conclusions.

We are also especially grateful for the assistance given to us by Major General Terry L. Gabreski. General Gabreski was the United States Air Forces in Europe (USAFE) Director of Logistics at Ramstein AB, Germany, during Operation Allied Force. General Gabreski also became the Air Force Forces (AFFOR) Director of Logistics, reporting to Lieutenant General Michael C. Short, who was both the Commander, Air Force Forces (COMAFFOR), and the Combined/Joint Forces Air Component Commander (C/JFACC) at Vicenza, Italy. In this capacity, she was responsible for planning and executing all combat support operations for U.S. aerospace forces in Europe. In our search for lessons that might impact the AEF combat support system of the future, General Gabreski provided free and open access to all involved under her command. We also want to thank her successor, Brigadier General Arthur J. Rooney, for his continued support of this work.

At the Air Staff, we thank Ms. Sue O'Neal, AF/ILX; Major General Scott C. Bergren and Mr. Grover Dunn, AF/ILM; Brigadier General Robert E. Mansfield, AF/ILS; Brigadier General Quentin L. Peterson, AF/ILT; and their staffs for their support and critique of this work. In particular, we thank Colonels Rod Boatright and John Gunselman for their time in reviewing and providing valuable feedback on earlier versions of this report. We also thank Lieutenant Colonel Patty Hunt, HQ USAF/CVAE, for her insights.

We have enjoyed support for our research from the Air Force's major commands involved with operations in Serbia. Major General Dennis G. Haines, ACC/XR, and Brigadier General Donald J. Wetekam, PACAF/LG, provided data and insights on operations in Serbia from their perspective.

In addition to Brigadier General Gabreski, many people at USAFE provided great support and access to data that were needed to accomplish our study. Colonel Dave Gillett, USAFE/ALG, made arrangements for us to interview the European Command/J-4 staff and ensured that we had access to the people and data needed for this report. Colonel Nettie Crawford, USAFE/LGX, coordinated our visits to USAFE bases and helped ensure that we met with the right people. Colonel Richard Kind, USAFE/LGM, provided keen insights into centralized intermediate repair facilities (CIRFs) and associated distribution challenges. Colonel Dennis Lami, USAFE/LGT, provided insights on distribution and in-transit visibility (ITV) results and issues. Colonel Susan Gallante, USAFE/LGS, provided direct access to her supply and fuels staff who participated in OAF. Colonel Russ Richardson, USAFE/LGW, provided insights on munitions challenges faced throughout operations in Serbia. Others on the USAFE staff who helped us include: from LGX, Major Dunbar, Major Humphrey, Major Mijares, Chief Master Sergeant Gardner, Captain Shearhouse, Mr. John Surovy, and Senior Master Sergeant Walker; from LGM, Major Lindsay, Major Freeman, Chief Master Sergeant Hauck, Chief Master Sergeant Hogue, Chief Master Sergeant McGuire, Chief Master Sergeant Schluetter, Chief Master Sergeant Vrahiotes, Senior Master Sergeant Garza, Senior Master Sergeant Haynes, and Senior Master Sergeant Mathers; from LGW, Lieutenant Colonel Maloney, Senior Master Sergeant Yeager, Senior Master Sergeant Alexander, Chief Master Sergeant Yates, Lieutenant Colonel Jobes, and Senior Master Sergeant Peters; from LGS, Lieutenant Colonel Fox, Captain

Jones, Mr. Young, Chief Master Sergeant Urban, Master Sergeant Berger, Master Sergeant Jones, and Master Sergeant Stubblefield; from LGT, Lieutenant Colonel Champeau, Captain Anderson, Captain Holmes, Lieutenant Hearn, Lieutenant Shaffer, Ms. Russell, Mr. Weismanten, and Chief Master Sergeant Franklin; from Air Mobility Operation Control Center, Senior Master Sergeant Miller, Senior Master Sergeant McAllister, and Master Sergeant Schwan; from USAF/XP, Major Nowland; and from 32 AOS, Major Kostelnik.

The logistics staffs at USAFE's two numbered Air Forces were greatly supportive in providing detailed information for this report. At 3rd Air Force, we thank Colonel Phil Miller, Lieutenant Colonel Fant, Captain Bulldis, Captain Kintz, Captain Pata, Lieutenant Porres, Senior Master Sergeant Nail, Master Sergeant Peartree, and Technical Sergeant Wilson. At 16th Air Force, we thank Colonel Pete Mooy, Lieutenant Colonel Meyer, Captain Milligan, and Captain Oliver.

Base-level personnel from the Logistics Groups at Aviano AB, Royal Air Force (RAF) Lakenheath, and Spangdahlem AB provided invaluable data for this report. They include: from Aviano AB, Chief Master Sergeant Holley, Senior Master Sergeant Taylor, and Senior Master Sergeant Gasper; from RAF Lakenheath, Colonel David Nakayama, Lieutenant Colonel Rhett Taylor, Chief Master Sergeant Levand, Chief Master Sergeant Mackey, Senior Master Sergeant Taylor, and Technical Sergeant Walls; and from Spangdahlem, Chief Master Sergeant Holas and Chief Master Sergeant Holmes. We also thank Senior Master Sergeant Murphy from the 104th Fighter Wing at Barnes Air National Guard base.

Major General Larry Lust, European Command/J-4, and his deputy, Colonel Dave Stringer, provided direct access to the European Command personnel and data, providing invaluable insight into the European Command's theater distribution system. In the Pacific, Major General Phillip Mattox, Pacific Command/J-4, and Major General Wade McManus, U.S. Forces, Korea/J-4, provided access to the their staffs so that comparisons among joint organizations and procedures could be made. The U.S. Forces, Korea staff, Colonel John Brown and Major Steve Lawlor, provided valuable data for this report. Colonel Ed Modlin, 7th Air Force/LG, and Lieutenant Colonel

l   Supporting Air and Space Expeditionary Forces

Wonzie Gardner, 607ASUS/CC, and their staffs provided comparison data that were used in this report.

At AFMC, Colonel Gary McCoy, AFMC/ALG, and Mr Curt Neumann, AFMC/XPS, helped in providing data on Air Logistics Center (ALC) logistics response times. At OO-ALC, Mr Maurice Carter helped explain current algorithms implemented at each ALC.

Our research has been a team effort with the Air Force Logistics Management Agency (AFLMA); the support of the AFLMA has been critical to the conduct of this research. We wish especially to thank Colonel Richard Bereit and Colonel Ronne Mercer, AFLMA Commanders during the period of this research. We also thank Lieutenant Colonel Mark McConnell and Lieutenant Colonel Steve Purtle, AFLMA/LGM, and Lieutenant Colonel Jeff Neuber, AFLMA/LGX, for their support. We especially wish to thank Major Donald Hinton, AFLMA/LGX, for his assistance in collecting data for this analysis.

We thank our Air Staff action officers, Colonel Frank Gorman, AF/ILXX, on the OAF effort and Colonel Connie Morrow, AF/ILGX, on OEF for their encouragement and support.

Finally, at RAND, we enhanced our analysis through the knowledge and support of many of our colleagues, especially Mahyar Amouzegar, Master Sergeant Les Dishman, Amanda Geller, Jim Leftwich, Patrick Mills, Marc Robbins, Amatzia Feinberg, Eric Peltz, and Bob Roll. We also thank RAND colleagues Walt Perry and David Oaks for providing thoughtful reviews and critiques of our work. For her detailed editorial review, which made this a better report, we thank Marian Branch. We thank Hy Shulman, Tim Ramey, Lionel Galway, Kip Miller, and Susan Bohandy, all of whom helped with reviews of this analysis. We thank Bob Wolff, Suzanne Gehri, and Dave George for their thoughtful review and critique of this work.

A-4	Logistics Directorate of the Air Component Staff
A/C	aircraft
ACC	Air Combat Command
ACC/CE	Air Combat Command/Civil Engineer
ACC/LG	Air Combat Command/Logistics Group
ACP	Ammunition Control Point
ACS	Agile Combat Support
AEF	Air and Space Expeditionary Force
AEFC	Aerospace Expeditionary Force Center
AETF	Air and Space Expeditionary Task Force
AF/CVAX	Task Force Enduring Look
AF/IL	Deputy Chief of Staff for Installations and Logistics
AF/XO	Deputy Chief of Staff for Air and Space Operations
AF/XOX	Deputy Chief of Staff for Air and Space Operations, Director of Operational Plans

AFCAP	Air Force Contract Augmentation Program
AFFOR	Air Force Forces
AFLMA	Air Force Logistics Management Agency
AFMC	Air Force Materiel Command
AFSC	Air Force Specialty Code
AFSOC	Air Force Special Operations Command
AIS	Avionics Intermediate-Maintenance Shop
AMC	Air Mobility Command
AMD	Air Mobility Division
ANG	Air National Guard
AOC	Air Operations Center
AOG	Air Operations Group
AOR	area of responsibility
APF	afloat prepositioning fleet
APS	Afloat Prepositioned Ship
ART	Active Reserve Technicians
ASETF-NA/CC	Commander, Air and Space Expeditionary Task Force–Noble Anvil
ATO	Air Tasking Order
BCAT	Beddown Capability Assessment Tool
C2	Command and Control
C2ISR	command and control intelligence, surveillance, and reconnaissance
CAOC	Combined Air Operations Center
CAP	Civil Air Patrol

CAT	Contingency Action Team
CC	combatant commander
CCP	Commodity Control Point
CE	Civil Engineer
CENTAF	U.S. Air Forces, Central Command
CHPMSK	Contingency High Priority Mission Support Kits
CIRF	centralized intermediate repair facility
COMAFFOR	Commander, Air Force Forces
COMJTF-NA	Commander, Joint Task Force Noble Anvil
COMJTF-OEF	Commander, Joint Task Force Operation Enduring Freedom
COMUSAFE	Commander, U.S. Air Forces, Europe
CONOPs	Concept of Operations
CONUS	continental United States
CS	combat support
CSC	Contingency Support Center
CSC2	combat support execution planning and control
CSL	CONUS support location
CV	deputy commander
CWT	Customer Wait Time
DCC	Deployment Control Center
DGATES	Deployable Global Air Transporation Execution System

DIRMOBFOR	Director of Mobility Forces
DLA	Defense Logistics Agency
DoD	Department of Defense
DWX	Deployable Major Regional Conflict/SS
DXS	Deployable SS
ECM	electronic countermeasure
ECS	Expandable Common-Use Shelter
EETL	Estimated Extended Tour Length
EUCOM	European Command
EW	Electronic Warfare
FAO	Foreign Area Officer
FOC	Fully Operationally Capable
FOL	forward operating location
FSL	forward support location
GATES	Global Air Transporation Exectuion System
GPS	Global Positioning System
HD/LD	high-demand/low-density
HF	Harvest Falcon
HIT	High Impact Target
HUMRO	humanitarian relief operation
IL	Installations and Logistics
IOC	initially operationally capable
ISR	Intelligence, Surveillance, and Reconnaissance
ITV	in-transit visibility

JAOP	Joint Air and Space Operations Plan
JCS	Joint Chiefs of Staff
JDAM	Joint Direct Attack Munition
JEIM	Jet Engine Intermediate Maintenance
JFACC	Joint Force Air Component Commander
JMC	Joint Movement Center
JOPES	Joint Operations Planning and Execution System
JPO	Joint Petroleum Office
JTF NA	Joint Task Force Noble Anvil
LANTIRN	Low Altitude Navigation and Targeting Infrared for Night
LG	Logistics
LRO	Logistics Readiness Officer
LRU	Line Replaceable Unit
LSO	Logistics Support Office
MAAP	master air attack plan
MAJCOM	major command
MANFOR	Manpower Force Requirement
MILAIR	Military Airlift
MOE	measure of effectiveness
MOG	maximum-on-ground
MRC	major regional conflict
MX	Maintenance

NAF	numbered Air Force
NORAD	North American Air Defense Command
OAF	Operation Allied Force
OEF	Operation Enduring Freedom
OSC	Operational Support Center
PAA	Primary Aircraft Authorized
PAF	Project AIR FORCE
PDC	power distribution center
PGM	precision-guided munition
POL	Petroleum, oil, and lubricants
POM	Program Objective Memorandum
PSAB	Prince Sultan Air Base
ROE	rule of engagement
RSS	Regional Supply Squadron
SDMI	Strategic Distribution Management Initiative
SITREP	situation report
SOF	Special Operations Forces
SSC	small-scale contingency
STAMP	standard air munitions package
STAR	standard air route
STEP	Survey Tool for Employment Planning
TAV	Total Asset Visibility
TDS	theater distribution system
TFEL	Task Force Enduring Look

TPFDD	Time Phased Force and Deployment Data
USAFE	U.S. Air Forces, Europe
USCOMCENTCOM	Commander, Central Command
USCOMEUCOM	Commander, European Command
USTRANSCOM	U.S. Transportation Command
UTASC	USAFE Theater Aerospace Support Center
UTC	unit type code
WCMD	Wind Corrected Munitions Dispenser
WOC	Wing Operations Center
WRM	war reserve materiel
WWX	World Wide Express

Chapter One

INTRODUCTION

The Air and Space Expeditionary Force (AEF) concept—substituting speed of deployment and employment for presence—was developed to allow the Air Force to respond quickly, with a tailored, sustainable force, to any national security issue. Since 1997, RAND Project AIR FORCE and the Air Force Logistics Management Agency have studied options for configuring a future Agile Combat Support (ACS) system that would enable AEF goals to be achieved.

BACKGROUND OF THE AGILE COMBAT SUPPORT SYSTEM

As defined in Tripp (2000), AEF operational goals are to

- rapidly configure support needed to achieve the desired operational effects
- quickly deploy both large and small tailored force packages with the capability to deliver substantial firepower anywhere in the world
- immediately employ these forces upon arrival
- smoothly shift from deployment to operational sustainment
- meet the demands of small-scale contingencies and peacekeeping commitments while maintaining readiness for potential contingencies outlined in defense guidance.

Two earlier RAND studies (Galway et al., 2000; Tripp et al., 1999) present the framework for an ACS system to support the AEF concept:

- A combat support execution planning and control (CSC2) system to assess, organize, and direct combat support[1] activities, meet operational requirements, and be responsive to rapidly changing circumstances. The CSC2 capability would help combat support personnel

  — Estimate combat support resource requirements and process performances needed to achieve the desired operational effects for the specific scenario.

  — Configure supply chains for deployment and sustainment, including the military and commercial transportation needed to meet deployment and sustainment needs.

  — Establish control parameters (for example, goals, maximum or minimum requirements) for the performance of various combat support processes required to meet specific operational needs.

  — Track actual combat support performance against control parameters.

  — Signal when a process is outside accepted control parameters so that plans can be developed to get the process back within control limits.

- A quickly configured and responsive distribution network to connect forward operating locations (FOLs), forward support locations (FSLs), and continental United States (CONUS) support locations (CSLs)

- A network of FOLs resourced to support varying deployment/ employment timelines

- A network of FSLs configured outside CONUS to provide storage capabilities for heavy war reserve materiel (WRM), such as munitions and tents, and selected maintenance capabilities, such as centralized intermediate repair facilities (CIRFs) that service jet engines of units deployed to FOLs. FSLs could be collocated with transportation hubs.

---

[1] In this report, the term *combat support* is defined as anything other than the actual flying operation. Combat support consists of civil engineering, communications, security forces, maintenance, service, munitions, etc.

- A network of CSLs, including Air Force depots, CIRFs, and contractor support facilities. As with FSLs, a variety of different activities may be set up at major Air Force bases, convenient civilian transportation hubs, or Air Force or other defense repair depots (Tripp et al., 1999).

Figure 1.1 is a notional representation of how the basic ACS concepts are being integrated to form a global ACS network that can enable AEF operational goals to be achieved across a wide variety of scenarios.

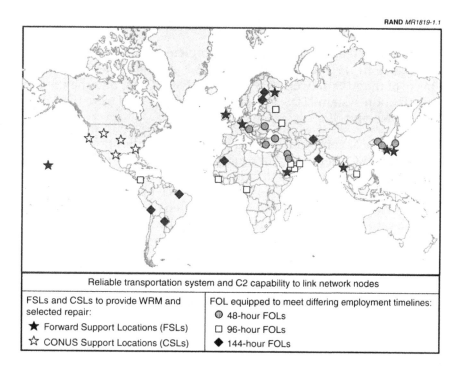

RAND *MR1819-1.1*

Reliable transportation system and C2 capability to link network nodes	
FSLs and CSLs to provide WRM and selected repair: ★ Forward Support Locations (FSLs) ☆ CONUS Support Locations (CSLs)	FOL equipped to meet differing employment timelines: ◉ 48-hour FOLs □ 96-hour FOLs ◆ 144-hour FOLs

Figure 1.1—Conceptual Global ACS Network

THE ACS NETWORK IN JOINT TASK FORCE NOBLE ANVIL AND OPERATION ENDURING FREEDOM

The ACS network was designed to support a wide variety of operational scenarios, from small-scale contingencies to major regional conflicts. In this report, we look at two recent U.S. military operations: Operation Allied Force (OAF), in Serbia in 1999, and Operation Enduring Freedom (OEF), in Afghanistan in 2001.[2] The organization overseeing the U.S. participation in OAF was named Joint Task Force Noble Anvil (JTF NA).[3] These two operations differed significantly; therefore, their experiences in using the ACS network concepts are of substantial interest to ACS concept developers.

Elements of the ACS network were implemented to enable operations in both JTF NA and OEF. Figure 1.2 illustrates the different aspects of the ACS network (CSC2, FOLs, FSLs, CSLs, and the theater distribution system [TDS]) and how and where they were used during JTF NA and OEF.

Because JTF NA and OEF differed in many ways, they provide a good test for the ACS network. For example, both contingency operations provided the opportunity to study how CSC2 concepts were implemented. In both contingencies, the Air Force played a major role in the development of TDS. FOLs were opened during both operations on sites that had not previously been used by the Air Force for exercises or in operations. These experiences provide insights on the ability of the Air Force to project power quickly to unknown FOLs.

Both operations used FSLs extensively—storage and maintenance FSLs and WRM, or supply, FSLs, including FSLs afloat in the form of the afloat prepositioning fleet (APF)—providing insight into how the global network of FSLs/CSLs can be improved for future operations.

---

[2]U.S. Air Force missions patrolling the airspace over the United States, named Operation Noble Eagle, took place at the same time as OEF and had significant combat support implications; however, those activities are not discussed in this report.

[3]Joint Task Force Noble Anvil was the organization overseeing all U.S. forces involved in Operation Allied Force. This report concentrates on Air Force operations conducted by Joint Task Force Noble Anvil.

RAND *MR1819-1.2*

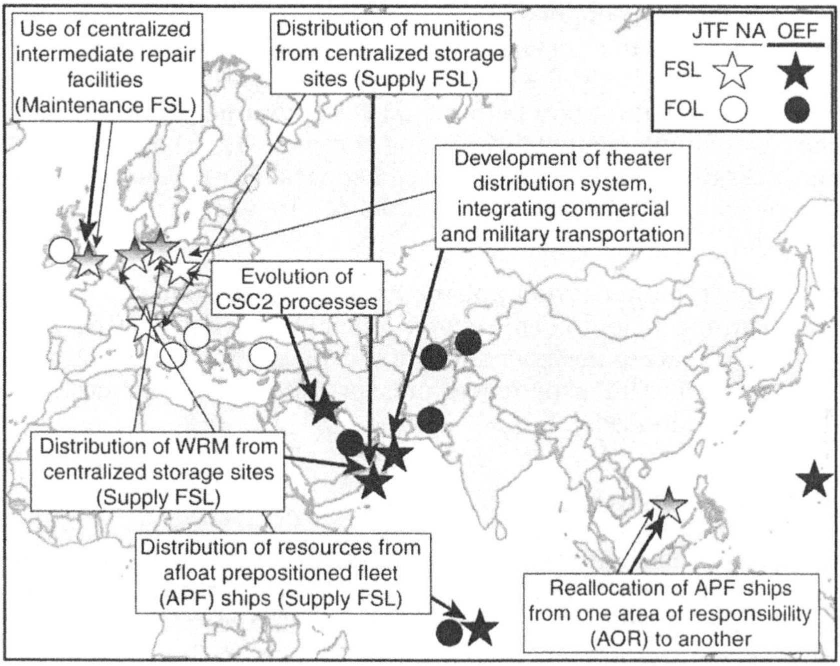

Figure 1.2—ACS Network as Implemented During JTF NA and OEF

Some of the ACS concepts—establishing effective CSC2 and rapidly configuring and developing a TDS to meet specific contingency operational needs—either were not fully developed or were not understood by the combat support and operational community before the contingencies began. Implementing the CSC2 and TDS concepts posed problems in JTF NA, and they continued to pose problems in OEF.

With other concepts, such as maintenance FSLs, much was learned during JTF NA, and the knowledge was successfully transferred to personnel engaged in OEF. JTF NA and OEF offered opportunities to examine the implementation of elements of the global ACS system in a contingency environment.

In 2000, Project AIR FORCE helped evaluate Agile Combat Support lessons from JTF NA. Some of the concepts and lessons learned from

JTF NA were implemented in support of OEF. This analysis allowed the opportunity to compare findings and implications from JTF NA and OEF. Specifically, the objectives of the analysis were to indicate how well combat support performed in OEF, examine how ACS concepts were implemented in OEF, and compare JTF NA and OEF experiences to determine similarities and applicability of lessons across experiences and whether some experiences are unique to particular scenarios.

This report analyzes the implementation of ACS concepts during these contingencies to determine whether (1) given the JTF NA and OEF experiences, the concepts should be modified and (2) the JTF NA and/or OEF experiences present additional ACS areas that should be addressed.

ANALYTIC APPROACH

Evaluating the performance of combat support in any operation raises the question: What measures should be used to judge the effectiveness of combat support activities? One measure could be whether constraints on combat support, such as shortages of resources, affected operations in an adverse way. For example, were sorties not launched because of combat support issues or did units not have the support needed to meet operational missions? From interviews with key participants and data collected from units engaged in JTF NA and OEF, we found that combat support was not a factor that inhibited operational mission performance during either contingency.

Since combat support met operational objectives, it may appear that little is to be learned from these experiences. However, we used a more stringent yardstick against which to measure combat support performance. Our study focuses on the performance of the JTF NA and OEF combat support systems relative to the ACS concepts that were developed to enable AEF goals. Thus, although official accounts of JTF NA indicate that "the logistics system . . . [and] logistics support was [considered a] success story during the air war" (Headquarters, USAF, 2000, pp. 45–46), we demonstrate that, during JTF NA, bottlenecks and analyses of scarce resources suggest several areas in which ACS could be improved to better achieve AEF goals.

Statements from both OEF operational and combat support leaders indicate that significant combat support issues were associated with OEF that raise serious concerns about supporting future contingencies. In the April 2002 issue of the *Armed Forces Journal International,* Lt Gen Michael E. Zettler, Deputy Chief of Staff for Installations and Logistics (AF/IL) during OEF, said "we made it up for Afghanistan as we went along . . . every one of those [missions] was an opportunity for failure . . . everything is needed" in that region of the world. This remark highlights issues associated with both combat support contingency planning and execution and the development of FOLs. Maj Gen Jeffrey B. Kohler, Director of Operational Plans for the Deputy Chief of Staff for Air and Space Operations (AF/XOX) during OEF, in a discussion in December 2001, expressed concern that in an operation where comparatively little "iron" was pushed forward, combat support resources were surprisingly strained. For combat support resources to be stressed in such an operation could presage difficulties in conducting other combat operations projecting larger force packages to austere areas around the world.

In this report, we use both empirical data from many sources and information from numerous interviews to analyze the development of CSC2, FOL preparation and development, FSL/CSL, reliable transportation, and resourcing both personnel and equipment to meet requirements of contingencies, rotations, and major regional conflicts (MRCs).

To gather information on the implementation of CSC2 concepts, we discussed the combat support chain of command during extensive field interviews with key participants, including the Commander, Air Force Forces (COMAFFOR) A-4, the COMAFFOR A-4 Rear, U.S. Air Forces, Central Command Deputy Commander (CENTAF/CV), U.S. Air Forces, Europe Logistics Group Commander (USAFE/LG), USAFE Theater Aerospace Support Center Commander (UTASC/CC), AF/IL, and their staffs. JTF NA data were also gathered from Air Force publications and from joint doctrine. We collected OEF data from situation reports (SITREPs), Air Force Combat Support Center daily briefings, CENTAF web sites, and Air Combat Command (ACC) Contingency Action Team (CAT) and Air Force Forces (AFFOR) A-4 Rear web sites.

JTF NA data on the development of FOLs were received from USAFE/LG, numerous CONUS-based Air Force organizations, and the Operation Allied Force Time Phased Force and Deployment Data (TPFDD). For OEF, we accessed Combat Support Center daily briefings, data from Task Force Enduring Look (TFEL),[4] including SITREPs, and the OEF TPFDD to gather information on FOL timelines. TFEL also provided data on OEF executive agency responsibility.

We conducted numerous field interviews during JTF NA to gather data on host-nation and contractor support. For OEF data, we interviewed CENTAF A-4 staff, the Director of Mobility Forces (DIRMOBFOR), and contractors on-site at CENTAF and in the area of responsibility (AOR).

In analyzing the amount of materiel moved and the method of transportation, we used the OAF and OEF TPFDDs and data provided by the contractor in the AOR. The Combat Support Center provided fuels data. CENTAF provided data on munitions and rations. Spares data were obtained from the Air Force Materiel Command (AFMC)/Logistics Support Office. To gather information about FSL throughput constraints and TDS, we interviewed the DIRMOBFOR, CENTAF staff, and the Deputy Director of the Joint Movement Center (JMC). Also interviewed were Air Mobility Command (AMC) and U.S. Transportation Command (USTRANSCOM) staff.

Information about how the Air Force is organized in the current AEF structure, including the specific construct for Agile Combat Support, was provided by the Aerospace Expeditionary Force Center (AEFC). Data on WRM were provided by CENTAF. AEFC, the Combat Support Center, and the Air Staff all provided personnel data about stressed career fields and the effect of resourcing the current AEF construct on those stressed career fields.

In this report, we compare specific lessons or experiences from OEF with those experiences documented during JTF NA to determine similarities, if any. We investigated whether JTF NA lessons were

---

[4]Formed in October 2001, Task Force Enduring Look is an Air Force–wide data collection, exploitation, documentation, and reporting effort for the air campaign against terrorism.

acted upon to improve the implementation of ACS concepts during OEF. When an experience was different or new during OEF, we assessed why and how it was different from the experiences during JTF NA.

Finally, we must point out that our analysis indicates how well combat support performed in JTF NA and OEF, not necessarily how well combat support *could* perform if demands were greater. Understanding experiences from this implementation could be of value for combat support and operations personnel who may be called upon to support future contingency operations.

ORGANIZATION OF THIS REPORT

Both JTF NA and OEF provide important opportunities to study how AEF ACS concepts were implemented during contingency operations. We begin with a description of the two contingencies and graphical comparisons, in Chapter Two. In the following five chapters, we look at five problem areas: CSC2 structure, in Chapter Three; FOL development, in Chapter Four; the use of FSLs and CSLs, in Chapter Five; the transportation system, in Chapter Six; and resourcing to meet current operational requirements, in Chapter Seven. Following the Conclusions, Chapter Eight, we provide three appendices: the future, or TO-BE, CSC2 operational architecture nodes and responsibilities, in Appendix A; a test of CSC2 in the centralized intermediate repair facility, in Appendix B; and a framework for assessing support capabilities, in Appendix C.

Chapter Two

AN OVERVIEW OF JTF NA AND OEF

Every military operation has its own unique characteristics. Therefore, neither the performance of the current support system nor the design of a future combat support system can be solely judged by or based on any one experience. That said, both JTF NA and OEF provide important experiences that warrant study for combat support operations. In this chapter, we discuss some characteristics of combat support during JTF NA and OEF to show similarities and differences between the two operations. We first compare the size and scope of these two recent U.S. Air Force operations.

OPERATIONS

For comparison, Figure 2.1 presents the size and scope of U.S. Air Force operations in Joint Task Force Noble Anvil and Operation Enduring Freedom. We compare JTF NA and OEF to illustrate the differences between what are both considered small-scale contingencies, although neither could be considered "small" in all aspects. The figure provides a quantitative comparison of the number of Air Force sorties flown, amount of munitions expended, number of beddown locations,[1] and number of Air Force personnel and aircraft deployed in these recent operations (Rosenthal, 2002; USAF, 437th Airlift Wing, 2002). By these measures, OEF could be considered a small combat operation compared with JTF NA.

---

[1]We use the term *beddown location* to refer to locations at which personnel and/or aircraft were based during operations.

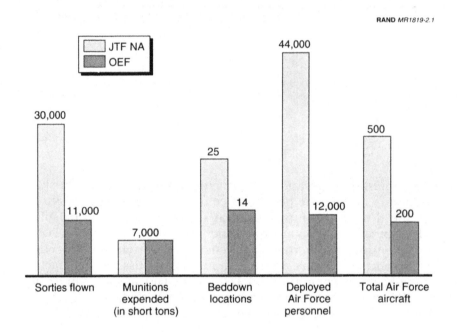

Figure 2.1—Size and Scope of JTF NA and OEF, for Comparison

During JTF NA, which took 78 days, 30,000 Air Force sorties were flown and 7,000 tons of munitions were expended.[2] In the first 100 days of combat in OEF (October 7, 2001, through January 14, 2002), fewer than half the number of JTF NA sorties were flown, but approximately the same amount of munitions was expended.

Overall, the munitions-expenditure rate was three times higher in OEF than in JTF NA, and many of these munitions were precision-guided. The use of precision-guided munitions was driven by the differences in rules of engagement. During JTF NA, the rules of engagement were extremely strict; the target sets required approval from the political leadership of the North Atlantic Treaty Organization member states. Of the munitions expended during JTF NA, 25 percent were precision-guided. In OEF, although not

─────────────

[2]Data were collected from USAFE and CONUS-based Air Force organizations and abstracted from the OAF TPFDD.

subject to the same level of political and coalition constraints as during JTF NA, the rules of engagement were driven by a desire to limit collateral damage. Precision-guided munitions accounted for approximately 46 percent of the munitions expended during OEF, which put a strain on the supply of these weapons.

Combat sorties do not tell the whole story. Of the 11,000 sorties flown in support of OEF, only approximately 3,400 were combat, command and control intelligence, surveillance, and reconnaissance (C2ISR), or Special Operations Forces (SOF)/combat support and rescue sorties. The Air Force conducted nearly 2,000 strategic lift missions, moving 83,000 tons of personnel and materiel on behalf of all the U.S. services. To complete the movement of personnel and materiel, 24 tactical airlift aircraft conducted 2,700 tactical lift sorties. Creating the air bridge[3] involved nearly 1,300 missions. The 49 aircraft performing tanking operations in the AOR flew approximately 4,700 sorties (Rosenthal, 2002; USTRANSCOM, 2002).[4]

Fewer Air Force aircraft were engaged in OEF (200) than during JTF NA (roughly 500). JTF NA involved a large fighter force, consisting mainly of USAFE-based forces; CONUS forces provided some augmentation, and there was limited naval involvement. OEF involved large SOF and naval participation, but limited fighter forces. JTF NA also had significant participation from coalition forces (as part of Operation Allied Force) (Lambeth, 2001).

Tankers supported operations in JTF NA; however, the distance to Afghanistan from carriers and land bases meant that OEF operations relied heavily on tankers for bomber and naval fighter operations (see Figure 1.2). The flight distance also affected maintenance, causing extensive maintenance to be performed on both the combat aircraft and the tankers. In support of SOF in-country operations, the Air Force Special Operations Command (AFSOC) MC-130 aircraft served as tankers.

---

[3]Established to move personnel and equipment to the area of conflict, an *air bridge* has three components: global reach laydown, strategic lift, and aerial refueling—everything necessary to move materiel to the AOR.

[4]Also see the Air Force news web site and the CNN web site.

The establishment of the air bridge during OEF required extensive combat support resources. The deployment of aircraft and personnel and the flow of cargo to the AOR required the deployment of ground crews, tanker aircraft, and airlift crews to intermediate locations. During OEF, two air bridges were established, one from the east coast of the United States and one from the west coast of the United States, to support operations in the Central Command AOR, which includes southern and central Asia. The Global Reach Laydown package[5] deployed nearly 1,800 personnel and 2,000 short tons of equipment to 28 different locations to support strategic airlift requirements (USAF AMC, 2002). In many locations, Tanker Airlift Control Elements were often the first personnel on the ground.

Initially, the tanker force was tasked to develop an air bridge to support strategic airflow. As the airlift airflow concepts developed, AMC dispersed maintenance and aircrews to nine locations to ensure the continuous movement of the strategic assets. Later in the operation, tankers were "loaned" to the theater commanders and supported both Air Force and Navy forces. In total, approximately 150 tanker aircraft and 200 tanker crews deployed to 16 different locations in support of OEF.

In addition to combat operations involving tankers, bombers, ISR (intelligence, surveillance, and reconnaissance) assets, and the air bridge, humanitarian relief operations (HUMROs) placed demands on the combat support system in both JTF NA and OEF. During OEF, from October 8, 2001, through December 21, 2001, over 2 million humanitarian daily rations were airdropped to Afghan civilians. Over 30,000 60-lb bags of wheat, 83,000 blankets, and hundreds of thousands of items of clothing were also delivered. C-17s flew over 200 sorties from Ramstein AB, Germany. These 15-hour, 6,000-mile round-trip missions were supported by tankers from Mildenhall, England, and fighters from Spangdahlem AB, Germany, and Lakenheath AB, England. The combat support for the fighters involved in HUMROs was forward-deployed to Incirlik, Turkey. In November, 6 KC-135 tankers and 200+ personnel deployed to Burgas, Bulgaria, to support the airdrop missions. In January 2002, following

---

[5]The Global Reach Laydown package provides ground support for AMC aircraft in the air bridge.

the cessation of airdrops, these forces redeployed (Martin, 2002; USAF, 437th Airlift Wing, 2002). USAFE supported both HUMRO and combat operations during OEF, an operation not in its AOR, adding to the complexity of the situation.

SUPPORT REQUIREMENTS

Some of the characteristics of OEF presented special challenges for combat support. The combination of short planning timelines and poor existing infrastructure created especially demanding requirements for combat support operations. Whereas the scale of OEF flight operations was comparatively small, support and beddown operations were large (see Figure 2.2).

One of the biggest support challenges was the compressed timeline of events. On September 14, 2001, following the attacks on September 11, 2001, the President activated Reservists. In an address on September 20, 2001, and in a subsequent letter on September 24,

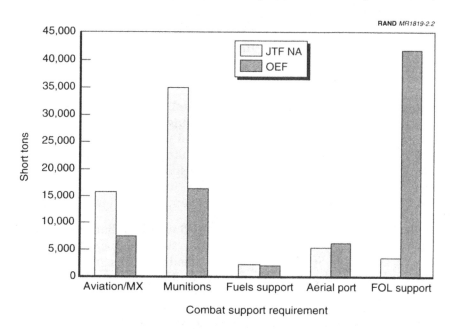

Figure 2.2—Combat Support Requirements During the First
100 Days of OEF

2001, the President informed Congress that he had ordered forces to deploy to the Central Command and Pacific Command AORs. The Central Command combatant commander issued an operational order on September 25, 2001, and operations commenced on October 7, 2001, less than one month after the terrorist attacks.

Limited planning time, combined with an immature theater infrastructure, resulted in the most challenging infrastructure development in recent history. The infrastructure was vulnerable to attack, unlike JTF NA, during which there was very little threat to the infrastructure. OEF presented sabotage opportunities for terrorists. The use of small arms and the potential use of handheld anti-aircraft weapons in proximity to the airfields required the heavy use of force-protection measures and personnel.

The aircraft involved in OEF were based in about half as many locations as those for JTF NA, which had 25 beddown locations. During OEF, approximately 12,000 people were deployed to approximately 14 separate locations[6] in the Central Command AOR. In addition, several sites in the European Command and the Pacific Command were augmented to support this operation, including Diego Garcia. Although Diego Garcia is in the Pacific Command AOR, its forces were placed under the operational control of Central Command during OEF.

Most JTF NA FOLs were well developed, as was the European Command theater infrastructure, including the transportation and supply infrastructures. Both commercial transportation and local industrial options were available, enabling the Air Force to use trucking, air, rail, and sea modes of transportation to meet deployment and resupply needs. In contrast, OEF depended upon opening numerous unknown, unanticipated, and unprepared FOLs.

Some FOLs in the Central Command AOR were familiar to Air Force planners and had been used previously to support Operation Desert Storm, Operation Southern Watch, or allied exercises. Prior to OEF, the Air Force had personnel, aircraft, and equipment deployed at

---

[6]The beddown locations changed during the course of OEF. The number listed here is an approximation. Because OEF was still ongoing at the time of this publication, the actual number and location of beddown locations are classified.

several bases in different countries in Southwest Asia. These forces were deployed to conduct Operation Southern Watch, the enforcement of the no-fly zone over southern Iraq. Operations over Afghanistan required that these forces be augmented. In some cases, existing bases were supplemented with additional personnel and infrastructure. Bases that once housed prepositioned equipment became transportation hubs or beddown locations for tankers, C2ISR assets, or bombers. A long distance from Afghanistan, these "known" FOLs were developed rather quickly to support operations, but because of their location, they required long flights and aerial refueling.

Before OEF, FOLs in the immediate area of the conflict were not as familiar to Air Force planners and were not prepared for immediate use. Many of the locations were fairly austere, some in bare-base environments lacking proper sanitation, and required significant infrastructure development. Consequently, high-demand/low-density (HD/LD) combat support personnel in such areas as combat communications, force protection, and civil engineering were required in significant numbers at most locations. In all, the number of personnel, aircraft, and beddown locations in the AOR to support OEF approximately tripled the Operation Southern Watch presence already in the AOR.[7]

Although OEF deployed fewer aircraft and involved fewer Air Force personnel than JTF NA, the number of items required for FOL support—for setting up and operating a base, such as tents, shower and lavatory facilities, generators, and lighting—was significantly larger than in JTF NA, as shown in Figure 2.2.[8]

Interestingly, fuels support requirements and aerial port requirements for both operations, measured in number of tons deployed, were approximately the same, even though OEF had fewer locations (14) than JTF NA (25).

_____

[7]See CNN web site, www.cnn.com/SPECIALS/2001/trade.center/military.map.html.

[8]FOL support data were extracted from TPFDD and obtained from the CENTAF contractor, DynCorp. DynCorp movements (other than air movement) are not shown in TPFDD.

JTF NA AND OEF IN PERSPECTIVE

Both JTF NA and OEF provide a glimpse of potential future conflicts. The Air Force should not base combat support system planning and decisions solely on any one operation. JTF NA had a long time to plan for an operation conducted from bases with good infrastructure. OEF was a small operation but was conducted on short notice in an immature theater. The next conflict could be similar to either one of these recent operations or it could present an entirely different scenario.

In May 2002, the Chief of Staff of the Air Force observed that "[the Air Force's] heightened tempo of operations is likely to continue at its current pace for the foreseeable future" (Jumper, 2002). The Air Force must be able to support the deployment of a large number of forces, either all at once in a major conflict or in an accumulation of a number of small-scale contingencies. Furthermore, it must be able to do so on short notice and in austere environments, particularly as the War on Terrorism continues around the world. The next chapter describes the experiences in planning and executing such support in JTF NA and OEF.

COMBAT SUPPORT EXECUTION PLANNING AND CONTROL

In this chapter, we address some of the key CSC2 experiences from JTF NA and OEF, along with the implications of those experiences for the future AEF. We begin by looking at a timeline for each operation and how the command and control structure developed as operations unfolded.

CSC2 NODES AND RESPONSIBILITIES IN JTF NA

CSC2 development in JTF NA was accomplished iteratively, over many months. Each new iteration took into consideration a new set of planning factors and operational requirements. Formal Operation Allied Force planning took place in late 1998. The first round of planning culminated in the creation of Joint Task Force Noble Anvil in January 1999. Figure 3.1 shows the timeline of developing CSC2 functions involved in JTF NA. Even with many months to plan, CSC2 development was ad hoc and did not follow doctrine.

Doctrine calls for the numbered Air Force (NAF) to transition to the wartime Air Force component role in times of conflict. Doctrine also calls for the augmentation of the NAF for reachback capability. During JTF NA, the Air Force chose to deviate from doctrinal guidelines and separated AFFOR and 'Joint Force Air Component Commander (JFACC) staffs into two separate locations (see

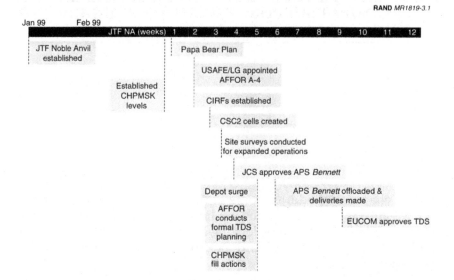

SOURCES: 32nd Air Operations Squadron, Ramstein AB, Germany, transcript of interview, January 31, 2000; Headquarters, USAFE Crisis Action Team, transcript of interview, February 2, 2000.

NOTE: APS=Afloat Prepositioned Ship; EUCOM=European Command.

Figure 3.1—JTF NA Operational and CSC2 Timeline

Figure 3.2).[1] Lt Gen Michael Short, 16th Air Force Commander, was selected to be the JFACC. The 16th Air Force A-4[2] was quickly overwhelmed by his responsibilities and looked to the major command (MAJCOM) component, USAFE, to provide support. At the beginning of JTF NA, USAFE had not yet clearly established roles and

---

[1]According to Air Force Doctrine Document 2, *Organization and Employment of Aerospace Power* (USAF, 1998), the NAF/CC and staff are delegated the COMAFFOR responsibilities. The NAF/CC acts as both the AFFOR, providing the "beds, beans, bullets, and Band-Aids" for Air Force forces, and the JFACC, overseeing the employment of all aerospace forces. Accordingly, the NAF staff is designated as both the AFFOR and the JFACC staffs, filling the Combined Air Operations Center (CAOC) and, when necessary, manning the JTF staff. On the basis of their doctrinal responsibilities, the NAF staff is the principal warfighting staff.

[2]The term *A-4* refers to the Logistics Directorate of the air component staff, which is responsible for logistics planning and execution for all Air Force activities in the area of responsibility.

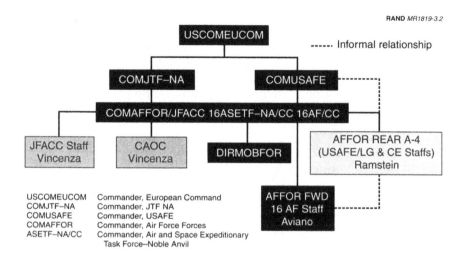

SOURCE: Jumper, n.d.

Figure 3.2—CSC2 Organizational Structure Implemented During JTF NA

responsibilities for executing these contingency responsibilities. The staff faced challenges in organizing to provide this support. They struggled to estimate their support needs and to present them to the European Command.[3] As JTF NA progressed, organizational roles and responsibilities evolved.

THE TO-BE OPERATIONAL ARCHITECTURE

CSC2 processes are not well documented in current Air Force or joint doctrine. As a result, both the operational and combat support communities have a limited understanding of the CSC2 process. This lack of understanding and an ad hoc organizational structure resulted in problems in combat support execution planning and control in JTF NA.

In response to the CSC2 issues discovered during operations in Serbia, AF/IL asked RAND Project AIR FORCE to study the current

---

[3]Information is based on extensive interviews and data collected from USAFE and CONUS-based Air Force organizations.

CSC2 operational architecture and develop a future, or "TO-BE," CSC2 operational architecture (Leftwich et al., 2002). Over the course of two years (2000 and 2001), RAND Project AIR FORCE documented the current processes, identified areas in need of change, and developed processes for a well-defined, closed-loop TO-BE CSC2 operational architecture incorporating the lessons learned during JTF NA.

More specifically, the TO-BE CSC2 operational architecture identifies the future CSC2 functions as including the ability to

- enable the combat support community to quickly estimate combat support requirements for force package options needed to achieve desired operational effects and assess the feasibility of operational and support plans

- quickly determine beddown capabilities, facilitate rapid TPFDD development, and configure a distribution network to meet employment timelines and resupply needs

- facilitate execution resupply planning and performance monitoring

- determine the effects, in-theater as well as globally, of allocating scarce resources to various combatant commanders

- indicate when combat support performance deviates from the desired state and implement replanning and/or get-well planning analysis to get the process back within control limits (Leftwich et al., 2002).

The TO-BE operational architecture was not completed until September 2001, just as OEF began. More details about the CSC2 TO-BE operational architecture can be found in Appendix A.

CSC2 NODES AND RESPONSIBILITIES IN OEF

The TO-BE architecture had not been vetted to senior leadership; therefore, it was not implemented during OEF. However, OEF offered another opportunity to examine the processes of the TO-BE architecture.

As in JTF NA, combat support command relationships during OEF did not follow doctrine. Doctrine calls for augmenting U.S. Air Forces, Central Command (CENTAF) A-4 personnel, deploying elements of the CENTAF A-4 forward, if forward operations are necessary, and establishing a reachback A-4 presence at the CENTAF Rear site at Shaw AFB, South Carolina. Instead of augmenting the NAF, the CENTAF A-4 and ACC/LG and ACC/CE established augmentation arrangements with ACC at Langley AFB, Virginia. Langley supported the A-4 reachback responsibilities of the CENTAF A-4, who went forward to Prince Sultan AB (PSAB), Saudi Arabia, to work with the COMAFFOR/JFACC (see Figure 3.3).

The speed with which the operation was executed is an important issue affecting CSC2 development in OEF. As the OEF operational planning timeline in Figure 3.4 shows, CSC2 command lines evolved over time, as did some forward and rear COMAFFOR A-4 functions. Not conveyed by the timeline chart is that the A-4 functions were not guided by doctrine or published guidance identifying the specific processes and functions that each CSC2 node would perform.

ACC/LG, ACC/CE, and CENTAF/A-4 recognized early that the CENTAF A-4 would need reachback support from ACC once the A-4 went forward. Reaching back to supporting MAJCOMs was a prac-

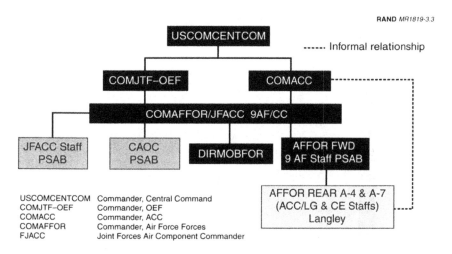

Figure 3.3—CSC2 Organizational Structure Implemented During OEF

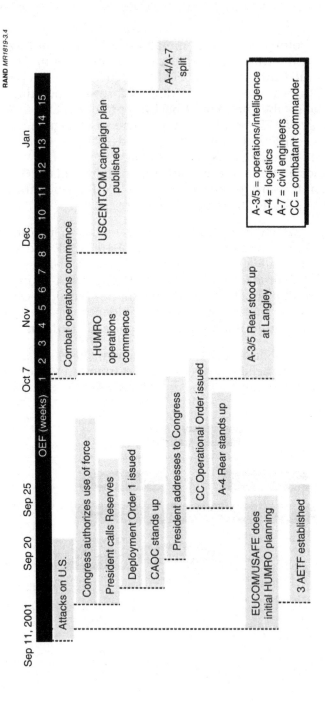

Figure 3.4—OEF Operational and CSC2 Timeline

tice used in recent military operations such as JTF NA, and now in OEF. However, in each case, this support was developed at the time of execution. As a result, no clearly defined or documented functions were identified for the A-4 Rear.

Each contingency used differing approaches. How reachback is implemented is left to the particular military personnel occupying the various CSC2 positions to work out; there is no playbook to aid them. Under these circumstances, then, support effectiveness may be a function more of the skills and experience of the personnel on hand at the time a contingency begins rather than of carefully organized support doctrine and policies.

ACC/LG and ACC/CE staffed an AFFOR A-4 Rear organization collocated with the Air Combat Command Crisis Action Team during OEF. This AFFOR A-4 Rear group understood that its job was not to support ACC/CC but to support the COMAFFOR as represented by the CENTAF A-4. The distinction is very important and was understood by all involved: they worked for the COMAFFOR. CENTAF A-4 personnel from Shaw AFB also augmented the AFFOR A-4 Rear operation.

In October/November, ACC/CE, the civil engineering staff, advocated that an A-7 be established forward to handle development of installations at the austere FOLs used during OEF. This position was established in February 2002. ACC/LG, the logistics group, concentrated on weapons systems sustainment and support. The AFFOR A-4 in the rear at Langley AFB remained a combined function; senior civil engineering and logistics colonels alternated heading the senior position. They reported to their respective CE and LG home organizations for assistance in resolving issues that were raised to CENTAF Rear at Langley.

The physical location of the AFFOR A-staff is important. It should allow the staff to concentrate on A-4 responsibilities—systemwide combat support planning and execution—and not encourage too much attention to Combined Air Operations Center (CAOC) daily Air Tasking Order (ATO) production. The AFFOR A-4 Forward in OEF indicated that collocation provided easy access to the JFACC/COMAFFOR and the AOC. The A-4 also indicated that A-4 functions were kept distinct during OEF.

The AFFOR A-4 staff should concentrate on assessing support effectiveness of alternative deployment and employment concepts and on identifying constraints to A-3/5 staff and the Air Operations Center (AOC). Although it did not correspond to their doctrinal responsibilities—doctrine calls for the separation of the A-staff functions and the CAOC functions—the AFFOR A-4 Forward functions were performed in the CAOC. Although not technically a break in doctrine, duties and responsibilities could become confused with A-staff and CAOC personnel collocated. Traditionally, and according to doctrine, the CAOC is responsible for developing the ATO. The logistics contingent in the CAOC was therefore responsible for assessing resources needed to support the ATO and for assessing operational impacts of resource shortages.

These roles are similar to those indicated in the CSC2 TO-BE operational architecture, which states that the AFFOR A-4 Forward would perform, plan, and assess support needed to meet the needs of the air campaign. Many of the requirements for support would be established by the A-3/5, and the CAOC would develop the ATOs to execute the plan. The A-4 would work with the A-3/5 to develop the campaign plan and assess operational plans to determine feasibility and resource implications of alternatives (Leftwich et al., 2002), including beddown assessments. During OEF, the AFFOR A-4 staff spent the largest proportion of their time dealing with beddown issues.[4] Because of the rapid response needed to conduct beddown assessments, the limited access to some sites, and the advancement of reachback technology—the ability to conduct initial site-feasibility assessments remotely—CENTAF A-4 Rear conducted many initial feasibility assessments.

The entire A-staff did not work out the same support arrangements during OEF. The A-3/5 also defined roles on an ad hoc basis. While the A-4 Rear was established at ACC, the A-3/5 reached back to Shaw AFB—an arrangement that made it difficult to develop an integrated campaign plan. Problems arose in developing TPFDD inputs, since the Joint Operations Planning and Execution System (JOPES) input capability was in the A-3/5 function at Shaw, but the A-4 inputs came from the AOR and Langley AFB. The A-3/5 function moved from

_____
[4]Interview with CENTAF/A-4 LGX staff, September 2002.

Shaw to Langley in October to alleviate coordination problems. However, in November, the A-3/5 Rear function moved back to Shaw.

This disjointed reachback raises the question of whether reachback operations should be separated by functional responsibility or whether all A-staff functions need to be collocated in standing rear organizations that can serve more than one COMAFFOR, as presented in the CSC2 operational architecture. For instance, if some other NAF were to have a sizable operation at the same time that the 9th Air Force was engaged in OEF, reachback for the new operation could be conducted from ACC in a section of an Operations Support Center that could support that operation while another section continued support for OEF.

While CENTAF supported forward beddown for deploying OEF forces, operational units associated with AFSOC and AMC developed reachback capabilities to their parent commands for sustainment support, bypassing CENTAF for reachback assistance for some items while relying on CENTAF for sustainment support for other items. At times, the result was confusion about which command had responsibility for support. Either because of this confusion or because of a lack of doctrine to guide these activities, reachback support was developed on an ad hoc basis.

CSC2 in Support of OEF Humanitarian Missions

The conducting of cross-AOR operations was another issue adding to the complexity of CSC2 during OEF. While Central Command led combat operations under way in Afghanistan, the European Command was conducting HUMRO operations.

The initial planning for the HUMRO operations began on September 12, 2001. The first missions were conducted on October 8, 2001, again demonstrating the speed with which command and control processes had to be developed. In this case, the reachback was to an organization that had been established before OEF commenced. Therefore, the reachback was well organized and did not involve ad hoc development of CSC2 as in both JTF NA and OEF combat operations. Figure 3.5 shows the USAFE command relationships for managing the HUMROs.

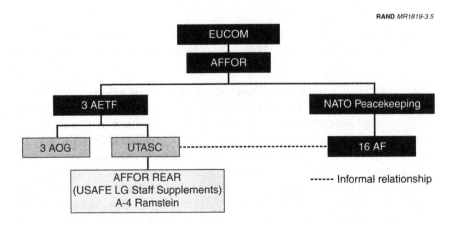

NOTE: AOG=Air Operations Group.

Figure 3.5—USAFE OEF COMAFFOR Command Relationships

Before OEF commenced, USAFE had established a USAFE Theater Aerospace Support Center (UTASC) to centrally manage day-to-day execution activities in the USAFE theater not assigned to a specific Air Force component command. UTASC developed the HUMRO operational plans and support concepts.

The 3 AETF was created to execute the plans that UTASC developed. USAFE conducted this mission with support from Army and coalition transportation personnel to supplement Air Force riggers[5] and perform other functions. To support this mission, USAFE provided resources from its own command in positioning forces and support personnel. These deployments were not recognized by the AEF Center in its AEF scheduling. This lack of recognition may have compromised the Air Force's ability to maintain AEF deployment policies.

Experiences from OEF have been incorporated into the TO-BE CSC2 operational architecture that has now been vetted and is in the process of being implemented.

---

[5]*Riggers* help load pallets and other cargo to make them air-transportable. They place the loose cargo on pallets, place nets over the loaded pallets, and balance the pallets to make them air-shippable.

Comparison of Actual and TO-BE CSC2 Nodal Responsibilities

During JTF NA, the USAFE/LG staff was organized into control cells to manage the combat support infrastructure, including the distribution system. With no policy to guide them, these control cells developed new reporting procedures to meet the needs of their customers. Innovative reporting and control processes were critical to the decisionmaking required to execute combat support as operations escalated.[6] These control cells resemble aspects of the CSC2 TO-BE operational architecture now being implemented.

In supporting Central Command's OEF responsibilities, COMAFFOR A-4 roles were developed on an ad hoc basis. In many ways, the roles resembled those described in the CSC2 TO-BE operational architecture. The UTASC A-4 performs functions that the CSC2 operational architecture indicates could be assigned to an Operational Support Center (OSC).

Both Headquarters Air Force and AFMC also developed CSC2 nodes similar to those in the CSC2 TO-BE operational architecture. These capabilities evolved during OEF without doctrine to guide their development.

At the Air Force level, the operational architecture calls for the Contingency Support Center (CSC) to monitor combat support requests for a particular contingency and assess the effects of those requests on the ability to support that and other contingencies. The existing Air Force Combat Support Center assumed many responsibilities of the future CSC, such as integrating multitheater requirements, identifying global resource constraints by commodity, conducting integrated assessments (base support), and recommending allocation actions for critical resources.

The Combat Support Center performed these functions. It intervened to allocate scarce resources to the AOR when those resources might have been designated to support other AORs and other poten-

---

[6]USAF (1998, pp. 32–35); Headquarters, USAF/CVAE Staff, transcript of interview, January 4, 2000; Headquarters, USAFE/LGT staff, transcript of interview, February 1, 2000.

tial contingencies. The Combat Support Center did the actual assessments for FOL support assets and relied on the supporting MAJCOMs to supply weapons system supportability assessments and provide an analysis of the effect of OEF operations on peacetime training and other potential contingency operations. The operational architecture calls for the CSC to conduct these weapons system and FOL support assessments. The Combat Support Center conducted weapons system assessment functions at the Air Staff; each MAJCOM conducted weapons system assessments, as well. In OEF, the Combat Support Center assessed the worldwide capability for FOL support and determined when the Air Force could provide support for other services—for example, Army Special Operations Forces—and made recommendations to the Joint Chiefs of Staff (JCS) accordingly.

Although it was able to support OEF, the Combat Support Center needs to be supplemented with analytic skills so that more-refined capabilities assessments can be made. Adding a capabilities assessment function and a limited number of combat support personnel with skills in quantitative methods should enhance this capability. This same team could support quantitative assessments needed to support the Program Objective Memorandum (POM) process during noncontingency operations (Hillestad, 2003).

Similarly, AFMC/LG assumed many of the responsibilities identified with a future spares Commodity Control Point (CCP) in the CSC2 TO-BE architecture, such as tracking spares shipments end to end, forecasting demands, and working more closely with customers and suppliers.

In 1994, the Logistics Support Office (LSO) was established at Headquarters AFMC. The LSO has an analysis section that monitors shipment pipelines in order to correct backorder problems, depot processing issues, and contract problems. The LSO also tracks the delivery times to various locations by various commercial and military modes of transportation. The LSO coordinates with AMC, commercial carriers, and personnel in the AOR to alleviate shipping problems. If a particular shipment was delayed, the LSO's CONUS Distribution Management Cell was empowered to reroute shipments. Delivery-time information was relayed to customers so that they would be able to make better decisions about transportation modes for future shipments. However, this information may not always reach the customer or be acted upon by the customer.

AFMC, along with its customer MAJCOMs, also set up a High Impact Target (HIT) list. Each MAJCOM identified a set of its most important repair parts for AFMC to monitor in the various Air Logistics Centers. This program is popular with the customer MAJCOMs, and AFMC has automated many of the processes associated with maintaining the list and gathering status reports. Other actions can be taken to strengthen these embryonic efforts to create the capabilities of the spares CCP as outlined in the CSC2 architecture.

INTEGRATED CLOSED-LOOP ASSESSMENT AND FEEDBACK CAPABILITIES

Another of the key concepts established in the CSC2 TO-BE operational architecture is a closed-loop assessment and feedback process,[7] a concept that has been well understood in operational planning and that has been the topic of operational planning doctrine for a long time (Boyd, 1987). This process can inform operations planners of how the performance of a particular combat support process affects operational capability. For example, in operations planning, it is standard procedure to conduct battle damage assessments and, if some targets have not been destroyed or rendered unusable, to modify the ATO to retarget.

Effective use of information feedback in combat support planning and control depends on two things: (1) reliable access to information and (2) a framework for measuring combat support process performances against "goals," or standards that are needed to achieve operational goals in the specific contingency operation. In OEF, as in JTF NA, not enough attention was given to combat support closed-loop feedback processes and to relating combat support process performance to operational goals. In fact, much of the feedback on the combat support process and many performance measures were incomplete. Many support decisions, such as capacity, manpower, and thresholds, were made without knowledge of how those deci-

---

[7]A *closed-loop process* takes the output and uses it as an input for the next iteration of the process.

sions might affect operational goals. Most combat support processes lacked the data-tracking capability to tie their actual levels of performance to performance levels that were planned for achieving specific levels of operational capability.

Many OEF support decisions were not based on operational needs or system status. OEF combat support response time goals were set arbitrarily or were based on historical performance, not on OEF operational requirements as they emerged. Progress has been made in developing information that could be used in combat support planning and control. For instance, several AFMC initiatives, the Strategic Distribution Management Initiative (SDMI), Total Asset Visibility (TAV), and other system improvements have developed information feeds of data to track current values of various transportation pipelines and the performance of other combat support processes.

The use of information systems has improved. Additional capability, such as providing better access to control information through automated system interfaces, is needed. The wider use of automated tools would enhance beddown assessments, and better links are needed between operational requirements and AFMC process performance and resource levels.

Not as much progress has been made on developing a combat support "closed-loop control framework," the essential elements and a detailed description of which can be found in the CSC2 TO-BE operational architecture (Leftwich et al., 2002). Support information that was tracked was not always used in decisionmaking. Information collected and analyzed by the AFMC LSO was transmitted to customers but was not always acted on. SDMI tracked current performance, but used broad, preconflict performance goals that did not keep up with evolving operational needs.

Appendix B presents a detailed explanation of the closed-loop feedback concept and discusses the details of how the concept was applied in managing CIRF operations during OEF. The CIRF CSC2 structure was modeled after the TO-BE process and demonstrates the feasibility of implementing the concept in a real contingency environment.

IMPLICATIONS

CSC2 is vital to any military operation. The planning of combat support needs to be integrated in the operational campaign planning process. The anticipated effects of alternative combat support strategies, tactics, and configurations need to be known to operations personnel when a plan is selected. In addition, once a jointly developed operations and combat support plan has been determined to be feasible and is capable of achieving the desired operational effects, a closed-loop feedback and control system needs to track the actual performance of combat support processes against planned values. When the system breaches control parameter limits, the CSC2 system needs to "signal" combat support personnel that corrective action is needed. The CSC2 TO-BE operational architecture (Appendix A) outlines how this planning and control could take place across the echelons of support and throughout the phases of operational campaigns. It could be used to guide CSC2 improvements.

The CSC2 operational architecture also specifies CSC2 nodes and associated responsibilities that are consistent with those that were developed on an ad hoc basis during both JTF NA and OEF. The CSC2 architecture specifies the broad responsibilities of the COMAFFOR A-4 Forward and Rear, the Combat Support Center, and Inventory Control Points. Many of the COMAFFOR A-4 functions, as well as those of the other nodes, can be performed by standing organizations. The UTASC, which was used to support USAFE OEF COMAFFOR responsibilities, is an example of a standing organization that is located in the rear, at the MAJCOM, and is referred to as an "Operational Support Center."

To implement the CSC2 operational architecture concepts requires changes in doctrine, education and training, organization, and systems, all of which should emphasize integrated operational-combat support closed-loop planning. Doctrinal changes have been initiated by AF/IL to begin the implementation of CSC2 processes and standing organizations discussed in the CSC2 operational architecture. This is a step in the right direction. But much more is needed.

In both JTF NA and OEF, staff augmentations were developed on an ad hoc basis. Which functions were to be performed forward and which were to be performed in the rear were not delineated clearly. However, based on experiences in JTF NA, organizational roles and responsibilities evolved more quickly during OEF. Command lines and responsibilities need to be clearly defined, especially the AFFOR Forward/Rear nodes, Inventory Control Point nodes, and the Air Staff node. Reachback command and control support needs to be thought through for all A-staff functions—something that has not been done. Education and training programs are needed to teach these concepts. In addition, performance feedback and control processes need to be improved and documented in policy. Decision support systems are needed to carry control information to combat support personnel so that significant deviations from planned performance can be corrected before operational effects are felt.

Chapter Four

FORWARD OPERATING LOCATIONS AND SITE PREPARATION

Just as the command organization varied in JTF NA and OEF, forward operating location development also varied. In this chapter, we discuss timelines for developing forward operating locations and combat support efforts associated with preparing for the deployment of forces during OEF, in comparison with JTF NA. We also discuss implications for future operations and the AEF concept.

FINDINGS

Our findings are in the following areas:

- FOL timelines
- Host-nation support and site surveys
- FOL development and construction
- Lift and flow issues
- Executive agency responsibilities in joint operations
- Contractor support.

JTF NA FOL Timelines

How long it took to develop FOLs in JTF NA varied greatly, as it did in OEF. Figure 4.1 compares the average amount of time it took to develop FOLs in JTF NA and in OEF.

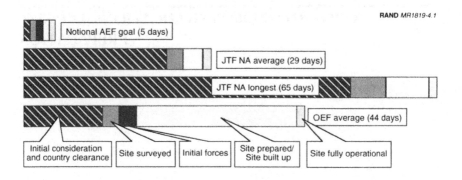

Figure 4.1—FOL Development in JTF NA

The top timeline in this figure illustrates a notional AEF goal: to deploy a force package to a known FOL and begin operations within a few days—for example, a five-day notional goal, with 72 hours for deployment and initial operations within 48 hours after arrival. The middle two timelines for the average time it took to develop an FOL during JTF NA (29 days) and the longest development time (65 days)[1] are vastly longer than the AEF goal. The average time it took to develop an FOL during OEF was 44 days.

The legend for the FOL site development timeline is located under the bottom timeline. The first section of each timeline represents the time it took to get country clearances once the site had been considered as a potential FOL. The second section represents the time it took to do an initial site survey. The third box shows the arrival of the initial airmen at each site. The next bar shows the time it took for site preparation and site development. The solid bar at the end of each timeline shows when the base was considered fully developed or *fully operationally capable* (FOC), which we define as a site with a full complement of base operating support in all functional areas, not necessarily a site that has received all assets listed in the TPFDD.[2]

---

[1]Information is based on extensive interviews and data collected from USAFE and CONUS-based Air Force organizations.

[2]*Fully operational* does not mean that all materiel and personnel needed at a site are present. It refers to resources that are in place or developed to meet mission needs. The definition of a fully operational site used here is roughly consistent with green

In comparing JTF NA and OEF, notice that during JTF NA the largest portion of time was spent obtaining country clearance and working the diplomatic issues to bed down forces at specific sites. During OEF, the largest portion of time was spent preparing the sites. An interesting note is the comparison of the total time needed to develop sites. In view of this information, the Air Force may wish to review bare-base beddown planning assumptions.

OEF FOL Timelines

As in JTF NA, some FOL locations used in OEF were adequately equipped to host U.S. forces with little buildup, whereas other sites were "bare bases" and required significant development. Diego Garcia was fully developed quickly, in approximately 17 days; Jacobabad took approximately 78 days to become FOC (HQ USAF, 2001a).

Figure 4.2 illustrates the development of four FOLs during OEF. Note that these timelines are relative to when each site was initially considered for deployment, not from one specific starting date, since site development began at different times for each site. The airplane symbol on each timeline shows when each base received aircraft and began conducting missions or was initially operationally capable

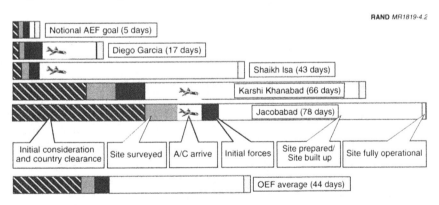

Figure 4.2—FOL Development Tmelines Varied in OEF

lights for all functional areas of the stoplight charts used in the HQ USAF (2001a) External IL Slides from the daily briefings.

(IOC). As shown in Figure 4.2, operations were conducted from FOLs well before the FOLs were fully developed.

Diego Garcia became a fully functioning FOL very quickly during OEF. Before OEF, the Air Force had determined that Diego Garcia would be a "bomber island"—one of several predetermined, prepared, forward operating locations for heavy bombers. As a result, the Air Force had extensive experience in deploying and operating out of Diego Garcia, and some bomber-island preparations were already under way. Thus, the FOL preparation time was less than that for other FOLs that were not as developed or whose capabilities were unknown to Air Force planners. Even though Diego Garcia was preplanned, it still took 17 days for the site to become FOC.

Jacobabad and Karshi Khanabad, in contrast, required extensive buildup of water purification, sanitation, and power generation, which had not been required at such preplanned FOLs as Diego Garcia. In addition, it took time at these "less-prepared" FOLs to set up force protection, repair deteriorating parking ramps, set up communications, build munitions pads, and develop large tent cities. Contracting for local area support also took time, placing demands on Air Force personnel capabilities. The FOL development timelines were much longer for these "unanticipated" FOLs.

JTF NA Host-Nation Support, Country Clearances, and Site Surveys

During JTF NA, agreements for host-nation support had an important delaying effect on FOL site-development timelines. Barriers to host-nation support were posed by both policy and politics. Policies for obtaining host-nation support and country clearance for conducting site surveys were not clearly delineated, and it was unclear whether the operational community or the logistics community would request the necessary host-nation support to conduct the site surveys.

During JTF NA, host nations were slow to grant country clearances, taking approximately 12 days to grant the clearance. Once clearance was received, there was no standardized site-survey checklist. The base support planning policy identifies a list of areas that should be addressed by the survey team, but a detailed, deliberate planning checklist is designed for teams that have time to conduct a lengthy survey. Once clearances were obtained, teams had very limited time—often only one day—in country. The detailed deliberate planning checklists were not suited for the type of survey conducted during JTF NA.

OEF Host-Nation Support, Country Clearances, and Site Surveys

During OEF, host-nation support and country clearance permissions were often delayed. With limited survey information available for some sites, CENTAF planners often had to rely on promised host-nation support rather than on detailed site surveys to accomplish initial site planning. In some cases, promised host-nation support was not delivered, or such support was slow to evolve.[3] The same issues encountered during JTF NA were faced again during OEF.

While obtaining clearance to enter a country was often difficult, obtaining access to specific sites during OEF was often more difficult. Specific access must be granted for site-survey teams to enter a potential FOL site. Often, survey teams were granted country access, but not desired site access, causing delays.[4] In some cases, access to potential beddown sites was denied.

Host-nation support changed as operations unfolded. For example, in late October/early November, Qatar, an important host nation, closed Camp Snoopy, an FOL already under development, which was located at Doha International Airport. Units were forced to relocate to Al Udeid Air Base outside Doha. Resources were consumed as units moved from an FOL under development to another site. These moves were outside the control of the Air Force and caused delays.

Air Force personnel routinely worked with U.S. Embassy personnel on such host-nation issues as diplomatic clearances. These embassy personnel are equipped to do their peacetime job, but they are not adequately staffed for wartime operations.

---

[3]Interview with CENTAF/A-4 LGX staff, September 2002.
[4]Interview with CENTAF/A-4 LGX staff, September 2002.

Difficulties involved in conducting site surveys caused part of the delay in developing FOLs during OEF. OEF required establishing FOLs in areas that had not before been considered as potential beddown locations. Survey data were not captured for all sites that were used in OEF, so services were caught with little or no site data for operations in the Afghanistan AOR. By itself, the use of unanticipated sites would have been a challenge. Further exacerbating the problem was the lack of a standardized site-survey process among Air Force commands, U.S. services, and allies—a lack that was also encountered during JTF NA.

The lack of survey standards and of a common site-survey tool complicated the survey process. Initial CENTAF and Central Command site surveys were conducted on an ad hoc basis. Multiple agencies did survey assessments, but not with uniform assessment procedures. Coalition teams, which did not always include civil engineers (Barthold, 2002, p. 2), were sent out with hard copies of a checklist without receiving training on how to conduct a survey using the checklist. Consequently, the checklists were returned incomplete, many in different formats, adding delays to deployment timelines. Incomplete information may account for the numerous incremental deployments of materials and personnel to complete the development of bases that could meet operational needs.

Politics also played a role in site surveys. Team composition was influenced by political sensitivities, which led to additional problems and delays. Air Force and coalition partners did not share common standards and expectations on the contents of site surveys. When coalition partners were in charge of conducting surveys with Air Force support, very different products were produced than when Air Force personnel led the process. Even Air Force–led surveys were nonstandard, in different formats and containing different information.

Within the Air Force and among the services, site-survey tools and techniques differed by command and mission type. Most combat support assessments were done quickly and manually and were not uniform in quality. Existing USTRANSCOM and AMC information on the AOR was not rapidly shared with Central Command and CENTAF.

Another challenge was the lack of a global site-survey database for information gathered during site surveys. Existing data were stored in many places, both physically and in many different databases, and the owning command controlled access to the data. Often information was not shared from one command to another or from one service to another. Information gathered during site surveys should be stored in a common database for use by all services.

Two site-survey tools are currently available to the services for use: Survey Tool for Employment Planning (STEP) and Beddown Capability Assessment Tool (BCAT). STEP/BCAT standardizes the data-collection approach and uses computer-generated templates to complete survey information. These systems were not used to collect data because site-survey personnel were not familiar with STEP/BCAT, did not wish to travel with classified equipment, or lacked the equipment and communication lines to update the Employment Knowledge Database[5] located at Maxwell AFB (Günter Annex), Montgomery, Alabama.

FOL Development and Construction

Once country clearances had been approved and sites surveyed, the austere conditions at several FOLs used in support of OEF required significant time for site development, as shown in Figure 4.3. During JTF NA, FOLs were developed in less-austere locations than in OEF. The infrastructure was more developed, so installation construction was not as extensive. The installation development of FOLs during OEF was the largest since Vietnam. The austere locations of many of the OEF FOLs required extensive engineering and development efforts. Existing buildings and facilities were unusable. The sites were not adequately developed for immediate Air Force use. In several cases, the Air Force deployed to very austere locations and commenced operations before FOLs were fully developed.

While not all living and working conditions were as poor as those shown in Figure 4.3, most sites were faced with some challenges. At some bases in the Southwest Asia AOR, the Air Force had preposi-

---

[5]The Employment Knowledge Database is a database that stores site surveys and base support plans for logistics planners' use.

RAND *MR1819-4.3*

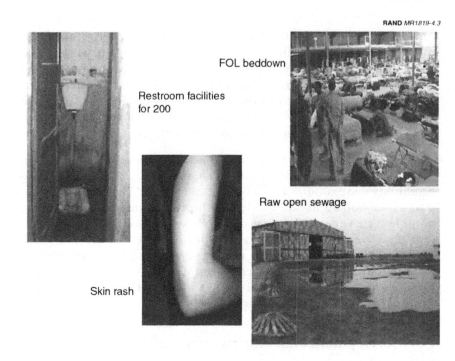

SOURCE: Pictures were provided by the Combat Support Center, Civil Engineer desk.

**Figure 4.3—"Creature Comforts" Suffered at Some FOLs
to Meet Operational Requirements**

tioned and existing facilities. Those bases were much better prepared for OEF deployment.[6] At sites not previously used in supporting Air Force deployments, most host-nation facilities required improvements.

To complete FOL construction, CE personnel were deployed to nine
locations in October 2001 in support of OEF. Of the 1,564 deployed
CE personnel, 848 were Active duty, 128 were Reserve, and 588 were
Guard. Construction projects included runway repair and ramp
construction, as well as construction of facilities for the airmen.

---

[6]For example, Seeb had a gymnasium and recreational center available within days of
the initial deployment.

In support of OEF, RED HORSE[7] teams of 500+ people worked on 77 projects valued at approximately $70 million. In October 2001, construction was planned or ongoing at seven sites: three existing Operation Southern Watch sites, three OEF sites, and one site awaiting construction. PRIME BEEF personnel were deployed to nine locations in October. By October 15, only 58 percent of the beddown locations were considered fully operational. By November 15, 66 percent were FOC; by December 18, 79 percent were FOC.[8]

In completing the construction work necessary during OEF, Civil Engineer (CE) resources, both personnel and equipment, became constrained. For example, FOL support assets were available in Jacobabad and ready for construction, yet no civil engineers were available to assemble them.[9] FOL development timelines were delayed because of stressed CE resources.

Lift and Flow Issues

CE resources were not the only assets constrained during OEF. Airlift was also limited. During JTF NA, the mature infrastructure helped ease issues surrounding the lift and flow of materiel. During OEF, strategic lift was limited and movement by air was prioritized daily according to combatant commander priorities, and those priorities may have caused some delay in moving materiel needed to supply the FOLs. During OEF, shipments of FOL base operating support materiel may have been delayed as other priorities, such as mail to the troops, took precedence in the AOR. The combatant commander consistently ranked mail to the troops the number-one priority during OEF.

There was also some confusion over the flow of assets needed to develop FOLs. For instance, site development at Manas, Kyrgyzstan, was initiated, then intentionally slowed to allow orderly development of the site. First the TPFDD flow of installation-development

---

[7]Rapid Engineer Deployable Heavy Operational Repair Squadron Engineer units make heavy repairs, upgrade airfields and facilities, and support deployment of weapons systems.

[8]Data provided by HQ USAF Combat Support Center, Civil Engineer desk.

[9]Interview with Maj Gen Richard Mentemeyer, October 2002.

assets was stopped, then the assets were reordered in the TPFDD, after which the flow began again. After the TPFDD adjustments, Manas development proceeded in an orderly fashion.[10] Manas is an example of a slow but successful FOL development.

Executive Agency Responsibilities in Joint Operations

In addition to issues regarding host-nation support, site surveys, and construction, executive agency responsibilities also shifted during OEF operations, affecting FOL support delivery and setup. During the beginning stages of OEF, executive agency responsibilities for FOL support were assigned to services by base and by country. In the Campaign Plan as of November 2001, out of the 14 sites developed in support of OEF, the Air Force had executive agency responsibility for only 50 percent, or seven of those sites. In reality, the Air Force assumed responsibility for 77 percent, or 11 of those sites. The Air Force assumed far more responsibility for FOL site development than was outlined in the November Campaign Plan.

In Karshi Khanabad, Special Operations Command Central had responsibility for FOL support until responsibility was turned over to the Army in late September. However, the Air Force was able to move forces in faster than the Army had planned for developing the site. As a result, the Air Force assumed the responsibility for FOL support and development for this and other FOLs. After initial site development actions, the Air Force and the Army shared development of Karshi Khanabad. Air Force PRIME BEEF civil engineers assisted in setting up Army FORCE PROVIDER assets in Karshi Khanabad.

Note that there is a cultural difference between Air Force and Army FOL support. Air Force units and their Army counterparts differ in their definition of what the standard of living should be in the AOR, Army standards being the more austere standards. This difference raises a question about joint-service FOLs: Since expectations are different, should the Air Force accept the development of shared Army/Air Force FOLs on a routine basis? Or should there be one accepted joint standard for all FOL support?

---

[10]Interview with HQ AF/ILGV, October 2002.

In addition to assuming some Army-assigned FOL development responsibilities, the Air Force provided FOL support to Special Operations Forces. Some SOF requirements were listed in the TPFDD and could be planned for; others were not and, therefore, were unplanned for. Even though the Air Force plans for SOF to use its resources, unknown requirements add stress to the management of WRM assets. Perhaps a certain percentage of FOL support resources should be reserved for SOF use.

Contractor Support

JTF NA relied on contractor support. The Air Force Contract Augmentation Program (AFCAP)[11] provided support for U.S. forces by moving facilities, conducting paving and facility evaluations, and providing heavy equipment in places such as Bosnia, Hungary, Turkey, and Italy (Wolff, 2000). AFCAP also established an FSL at Ramstein AB, Germany, from which it supported the construction of a small city of 17 kilometers of roads, 1,820 tents, 1,006 latrines, 270 water taps, 12 school areas, 44 bath houses, and 176 food-preparation areas in 51 days (Wolff, 2000).

During OEF, the development of FOLs was also aided by contractor support already on site in many locations. Contractors at WRM storage locations were able to shift from WRM maintenance to FOL site-preparation work at collocated FSL sites. Contractor support helped ease the installation-development workload for the military personnel in both JTF NA and OEF.

Contractors supported site preparation in five locations during OEF. They helped civil engineers establish Camp Snoopy, construct tent cities in two locations, set up fuel farms, and refuel aircraft until Air Force personnel arrived. They also operated power plants at several locations (DynCorp, n.d.). Contractors provided equipment, ground transportation, bottled water, furniture, facilities, cellular telephones, laundry services, and fuel. They worked 485,772 hours of overtime and catered 1,279,187 meals. Additional contractor personnel were hired, and existing personnel were reallocated to sup-

---

[11]AFCAP is a contract tool, available only during contingency response, to provide civil engineering and services support.

port the needs of OEF. See Table 4.1 for an example of the increased contractor workload during the first 100 days of OEF.

Although contractors have a contingency clause in their statement of work that allows this sort of additional support, the support provided by the contractor was extremely beneficial in meeting rapid FOL development and aiding in uninterrupted sustainment. However, one consequence of using contractors in this capacity was the reduced *outload capability*, the ability to move materiel out of the FSL, which is discussed in the next chapter.

IMPLICATIONS

Selection and development of FOLs play an important role in meeting the AEF goal. Many actions can be taken to decrease deployment times and reduce FOL preparation times as experienced in JTF NA and OEF.

Large amounts of time were expended in both JTF NA and OEF in gaining country clearances and specific FOL access. Engagement policies and programs to familiarize Air Force planners with facilities in countries that may be sites for future operations could potentially reduce country access time. In such programs as Partnership for Peace, military-to-military contact is encouraged, and exercises and deployments are conducted through which knowledge of FOLs can be gained. Participation in such programs could be valuable for helping speed deployments to important areas around the world and should be encouraged.

Table 4.1

Contractor Support Surged During OEF

	Average for OEF[a]	Average pre-OEF
Direct mission support tasks	162	10
Tons of air cargo moved	1,554	400
Truckloads moved	705	200
Total tons moved	9,331	3,000

[a]Average is calculated from October 2001 through January 2002 data.

Training some Air Force combat support officers in a fashion similar to the Army Foreign Area Officer (FAO), as an expert in the language, culture, and politics of the area, could produce some country and area specialists. FAOs are language specialists, understand the culture, and understand host-nation requirements/restrictions. Such FAO-like support officers could augment embassies in the early stages of a conflict and facilitate rapid country clearances, access, and host-nation support agreements.

Programs with engagement opportunities should be leveraged to develop site-survey data. The site survey processes in both JTF NA and OEF were ad hoc, and efforts to standardize site-survey approaches need to be made within Air Force commands, between services, and among U.S. allies. Perhaps standing site-survey teams should be established. The teams could then be trained to use standardized tools that result in standardized surveys. Technology exists today that can offer opportunities for improvement in the site-survey process.

Currently, 11 separate agencies maintain site-survey databases. The owning command has control of who may or may not have access to the site-survey information in its system. Perhaps the Air Force portal, or a classified version of the portal, could be used to coordinate a common database of site-survey information. Much more can be done to enhance site-survey capabilities within the Air Force and between U.S. allies and other services.

In some cases, sites in potential "hotbed areas" could be prepared in advance for rapid deployment. Where possible, a select number of future FOLs in these areas should be surveyed for capabilities. Goals could be established in each AOR for surveying potential sites for future Air Force use. Funds could be put aside for accomplishing these surveys.

Other opportunities exist for decreasing FOL preparation time, including leveraging contractor capabilities, where available, to assist civil engineers in developing FOLs once a contingency begins and augmenting embassy staffs with an area-expert such as an FAO to help gain country access and clearances in the early stages of contingencies, when staffs at embassies are often overwhelmed.

AFCAP and other contractor capabilities, such as WRM maintenance contractors at FSLs, can be leveraged to aid civil engineers in rapidly building up and then sustaining FOLs, as demonstrated in OEF. Although it may be desirable to have Air Force civil engineers complete the initial beddown planning and construction, capabilities to augment scarce Air Force personnel skills could be developed through these programs. Databases of contractor capabilities, similar to FOL site surveys, should be developed in areas in which potential conflicts may be likely.

Chapter Five

FORWARD SUPPORT LOCATION/CONUS SUPPORT LOCATION PREPARATION FOR MEETING UNCERTAIN FOL REQUIREMENTS

The ability to quickly link a global network of forward support locations (FSLs) and CONUS support locations (CSLs) to meet FOL deployment and sustainment needs is vital to every operation. In this chapter, we analyze JTF NA and OEF data to illustrate the importance of this global network. We also discuss some of the limitations of the current network in meeting operational needs.

FINDINGS

Our findings are in the following areas:

- FSLs as supply locations

- CSLs and C2

- Maintenance FSLs/CIRFs.

FSLs as Supply Locations

Combat support resources dominated the total movement in both JTF NA and OEF, as shown in Figure 5.1. An analysis of the JTF NA and OEF TPFDDs, as well as data provided by the CENTAF WRM contractor, DynCorp, for OEF, shows that during JTF NA, aviation units and their associated maintenance functions accounted for only 20 percent of the tonnage moved to FOLs; during OEF, they accounted for only 9 percent of the movement. Aerial port equipment

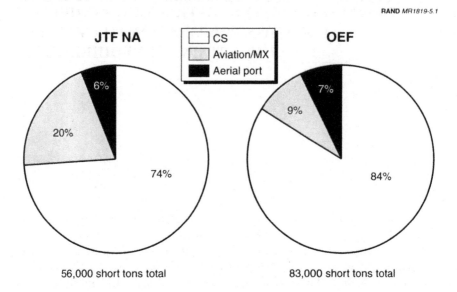

RAND *MR1819-5.1*

Figure 5.1—JTF NA and OEF FOL Materiel Movement

accounted for 6 percent of the movement during JTF NA and only 7 percent of the movement during OEF. The remaining 74 percent of movement for JTF NA and 84 percent of movement for OEF consisted of Air Force combat support resources.

An analysis of the combat support portion of the OEF TPFDD and DynCorp data shows that the largest portion, 65 percent of the requirement, is FOL support, as shown in Figure 5.2. The term *FOL support* is used to identify the *base operating support*—those resources that are required to set up and sustain a base. This resource category includes, but is not limited to, civil engineering equipment; WRM, including tentage, shower/shave, and water-purification systems; and vehicles. Also included in this category are the industrial and kitchen sets that round out the base support packages. During JTF NA, WRM was distributed from Sanem, Luxembourg, and CONUS. During OEF, WRM was distributed from Al Udeid Air Base, Qatar; BANZ Industrial Site, Manama, Bahrain; Thumrait Air Base, Seeb Air Base, and Masirah Island Air Base, Oman; Sanem, Luxembourg; and CONUS locations.

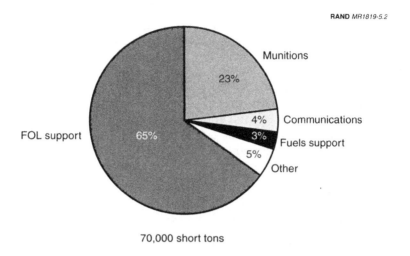

Figure 5.2—OEF Combat Support Requirements

Other combat support resources include munitions, communications, and fuels support, such as bladders and pumps. During OEF, munitions resources accounted for 23 percent of the total combat support requirements, whereas communications equipment accounted for only 4 percent. Fuels mobility support equipment—such as, bladders, hoses, pumps, not including fuel itself—made up 3 percent of the movement requirement. The remaining 5 percent is split between various combat support requirements. Note that these percentages include items listed in the TPFDD and those moved by DynCorp. Food, water, and fuel are not included in these figures.

During JTF NA, FSLs and CSLs satisfied approximately 76 percent of the combat support requirements. During OEF, FSLs satisfied the largest portion of the combat support requirement—approximately 64 percent (see Figure 5.3). Included in that 64 percent is the approximately 82 percent of the total FOL support resources needed during OEF, which was provided by FSLs. The other 16 percent of the FSLs' support was munitions related.

During OEF, CSLs also satisfied a portion of the FOL requirements, although a much smaller portion of the overall support: only approximately 11 percent. Of the combat support resources moved from

CONUS, only 13 percent was for FOL support. Most of the CONUS support, approximately 85 percent, was munitions related (see Figure 5.3).

Although FSLs provided the majority of the FOL total resource requirements during OEF, they did operate with some constraints. Many of the FSLs are located at or near places the Air Force intended to use as FOLs—for example, Al Udeid and Thumrait Air Bases. Moving bombers or tankers into a location while moving support equipment out created issues with ramp space and equipment utilization. The FOL ramp requirements limited FSL throughput. Workload and beddown requirements at these locations created conflicts. Some sites completely stopped the outload of equipment, the movement of equipment out of the FSL to sites at which the equipment was needed, while the contractor teams helped to build tent cities prior to force packages arriving at combined FSL/FOL locations.

Such contention for contractor resources could adversely affect deployment timelines to other FOLs at which outloads from the FSLs are needed to set up and begin operations, causing a domino effect. Additionally, ramp space that is consumed by aircraft operating from

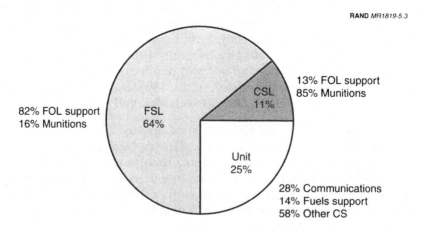

RAND MR1819-5.3

Figure 5.3—Combat Support Requirements Were Resourced Mainly from FSLs During OEF

the combined FSL/FOL site is not available for airlift aircraft to move equipment out of the FSL. Many locations found themselves short on material-handling equipment and qualified contractor personnel to operate the available equipment. Processes were not in place to shift from peacetime maintenance of WRM to the wartime distribution of those same assets. Also, the contractor had limited personnel who could complete all the paperwork (Bills of Lading) necessary for the shipments to transit numerous borders. However, even with these problems, FSLs did satisfy most FOL resource needs.

CSLs and C2

During JTF NA, constraints on CSL resources hindered CSL effectiveness. Specifically, backorders added substantial resupply time and variability during the conflict. Prioritization of supply resulted in an unequal readiness level in CONUS and across the rest of the Air Force. Although backorder rates improved, they remained high throughout Operation Allied Force.

CSLs were used more effectively during OEF. Because of JTF NA experiences, attention was given to creating better links between CSLs and the warfighters. To enhance CSL responsiveness to the warfighter, AFMC tasked the Logistics Support Office, Headquarters AFMC, to monitor shipment pipelines and track the delivery times to various locations by various commercial and military transportation modes. Delivery-time information was relayed to customers so that they could make better decisions about transportation modes for future shipments. AFMC/LSO also developed a web site with estimated shipping times and best methods for shipping to different locations. The web site was updated to show any anomalies in shipping due to customs problems or host-nation restrictions so that alternate routing could be used.

AFMC also created a HIT list. MAJCOMs identified their most-important repair parts for AFMC to monitor in the various Air Logistics Centers. This program is popular with the customer MAJCOMs. AFMC now automates many of the processes associated with maintaining the list and gathering status reports.

AFMC/LG assumed many of the responsibilities identified with a future spares commodity control point in the CSC2 architecture.

They created the CCP and specified wartime processes as outlined in the CSC2 operational architecture.[1]

Maintenance FSLs/CIRFs

During JTF NA, the use of centralized intermediate repair facilities (CIRFs), also called maintenance FSLs, was successful in meeting the warfighters' needs. Three existing USAFE FSLs were formally designated as CIRFs during JTF NA: RAF Lakenheath, England; Aviano AB, Italy; and Spangdahlem AB, Germany. RAF Mildenhall, England, was later developed as a CIRF to support tankers. JTF NA showed that preselection and resourcing of CIRFs can improve flexibility and reduce the deployment footprint.

As a result of successfully using CIRFs on an ad hoc basis during JTF NA, the Air Force developed and tested an official CIRF Concept of Operations (CONOPs) for supporting the AEF. The Air Force CIRF test began in September 2001. The CIRF CONOPs was adjusted to include support to OEF forces once those operations began.

CIRFs were established in USAFE to satisfy a range of intermediate repair operations for fighter units deployed to Operation Northern Watch and Operation Southern Watch/OEF (see Figure 5.4). The repair facilities at RAF Lakenheath were identified for F-15 line replaceable unit (LRU) repair, as well as for maintaining Low Altitude Navigation and Targeting Infrared for Night (LANTIRN) pods and F-100 engines. Spangdahlem Air Base in Germany was designated as the repair facility for ALQ-131 electronic countermeasure (ECM) pods and F-110 engines. Later, when a backlog developed at Lakenheath, USAFE added the LANTIRN repair facility at Aviano Air Base in Italy. The transportation segment that had been planned to support Operation Northern Watch and Operation Southern Watch was expanded to include the other OEF locations. Plans were developed to move items from forward bases to Operation Northern Watch/Operation Southern Watch locations for movement onward.[2]

---

[1]See Chapter Three for more details on the TO-BE CSC2 operational architecture.

[2]The theater distribution system will be covered in more detail in Chapter Six.

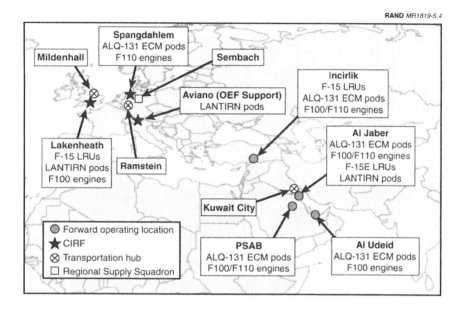

RAND MR1819-5.4

Figure 5.4—CIRFs Provided Maintenance Support for Fighters During OEF

The CIRFs reduced the deployment requirement during OEF, as shown in Figure 5.5. For example, the CIRF was able to support all Southwest Asia repair needs for ALQ-131 pods with the existing equipment at Spangdahlem and nine additional personnel. If this repair capability had been deployed forward, each deployed unit would have needed seven personnel and 13 short tons of support equipment. With five deployed units, these requirements would total 35 personnel and 65 short tons of equipment.

Similar savings were achieved for Jet Engine Intermediate Maintenance (JEIM), LANTIRN, and Avionics Intermediate-Maintenance Shop (AIS) personnel and equipment. These three CIRFs combined to save the deployment of 79 personnel and almost 123 short tons of support equipment during OEF that would have been required under a decentralized structure.

In addition to fighter CIRFs, B-52 phase maintenance was moved to a forward location to increase aircraft availability during OEF. B-52s

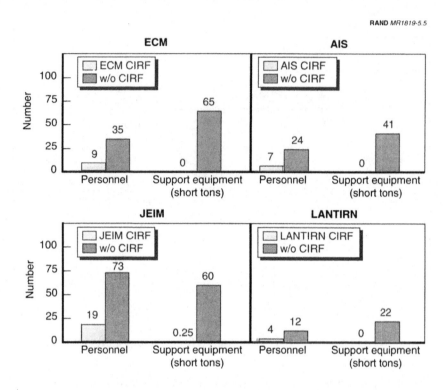

Figure 5.5—CIRFs Reduced the Southwest Asia/AOR Footprint

were initially returned to home station for phase maintenance inspections. In addition to the fuel consumed to make these return
trips to CONUS, many flying hours were consumed, and the extra
sorties required pilots to fly more hours. In late December 2001,
phase maintenance was moved forward to Andersen AB, Guam. To
support phase maintenance in Guam, 26 short tons of equipment
was moved initially. The first 70+ airmen arrived on December 21
and completed their first phase of maintenance five days later.
Approximately 40 phase inspections were completed at Guam, sav-
ing over 1,200 flying hours and reducing the tanker air-bridge requirements.

IMPLICATIONS

A review of JTF NA and OEF clearly points to the global nature of future conflicts. The current AEF force structure of light, lean, and lethal response forces depends heavily on FSLs. A global network of CSLs and FSLs with prepositioned WRM is necessary to meet AEF goals. Having to use austere FOLs and an immature theater infrastructure in OEF has illustrated the need for a "portfolio" of FSLs. Current progress in developing maintenance and supply FSLs is on track.

However, sustained senior leadership is needed to enhance the FSL portfolio for meeting uncertain contingencies and to balance risks and costs for the future. When developing a portfolio of FSLs to support different operational challenges, many options should be provided and available for use. Trade-offs between improving existing FSLs, which may enhance throughput and storage capacity, and capabilities that can be developed by investing in new FSLs in differing locations will need to be examined.

When considering whether to develop new FSLs or improve existing facilities, planners should pay attention to joint requirements, in particular beddown assets and the support for both aviation and ground SOF units. All services depend on prepositioned materiel to meet contingency requirements. Managing joint facilities to meet multiple service requirements may reduce operating costs. Likewise, information needs to be shared among services as well as with U.S. allies. If such arrangements are pursued, throughput required to meet all participants' needs must be considered explicitly.

Better linkages are needed to connect CSL spares workloads to warfighter needs. AFMC has developed several tools, such as the HIT list, that should be enhanced and used in future operations. FSL and CSL throughput capability may constrain deployment speed and must be considered during planning processes.

Chapter Six

RELIABLE TRANSPORTATION TO MEET FOL NEEDS

The movement of personnel and equipment is vital to an operation's success. Without a reliable transportation system, deployment can be delayed and sustainment can be hindered. In this chapter, we discuss transportation and movement experiences in JTF NA and OEF.

FINDINGS

Our findings are in the following areas:

- Movement by commodity
    - Fuel
    - Munitions
    - FOL Support Assets
    - Spares.
- Management of the theater distribution system (TDS)
    - TDS responsibility and organization
    - In-transit visibility (ITV) and communications
    - Using feedback to close the loop.

Movement by Commodity

During both JTF NA and OEF, the TDS movement requirements were dominated by such commodities as fuel,[1] munitions,[2] FOL support, rations,[3] and spares.[4] These are bulky commodities; moving them requires large transportation capacity. Figure 6.1 shows the movement requirements of these items during the first 100 days of OEF.

Fuels by far dominated movement requirements. Movement of the non-fuels requirements in support of OEF is broken out in Figure 6.2. Munitions and FOL support, which accounted for a large portion of the movement, are discussed in this chapter. Even though spares accounted for a small portion (only approximately 10,000 short tons), they are critical to weapons system support; we also address their movement in this chapter.

RAND *MR1819-6.1*

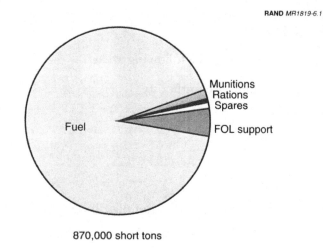

870,000 short tons

Figure 6.1—OEF Commodity Movement

---

[1]Fuels data were provided by HQ USAF Combat Support Center, Fuels Consumption Log, July 2002.

[2]Munitions data source: USAF, 437th Airlift Wing (2002).

[3]Rations data were provided by CENTAF/A-4 LGV.

[4]Spares data were provided by AFMC/LSO-LOT.

RAND *MR1819-6.2*

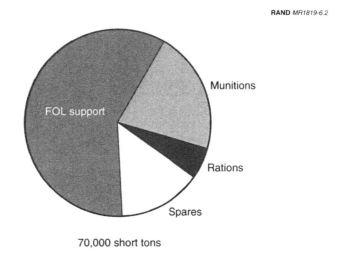

70,000 short tons

Figure 6.2—Commodities Other Than Fuel Moved in Support of OEF

The Air Force needs to move large amounts of materiel to initiate and sustain combat operations, and it must use many modes of transportation and different commodity supply chains to do so. Figure 6.3 illustrates the modes of transportation used during OEF to move combat support materiel only, not aviation, maintenance, or aerial port materiel.[5] Intratheater airlift handled approximately 15,400 short tons.[6] Approximately 11,000 short tons were moved by sea,[7] 34,000 short tons were moved by land,[8] and 1,200 short tons were moved by intertheater airlift.[9]

We now take a more detailed look at several of the commodities.

_____

[5]The 67,000 short tons in Figure 6.3 does not include the 3,000 short tons of rations moved in support of OEF.

[6]Data were abstracted from OEF TPFDD.

[7]Data were provided by Ammunition Control Point (ACP).

[8]Data were provided by DynCorp.

[9]Data were abstracted from OEF TPFDD.

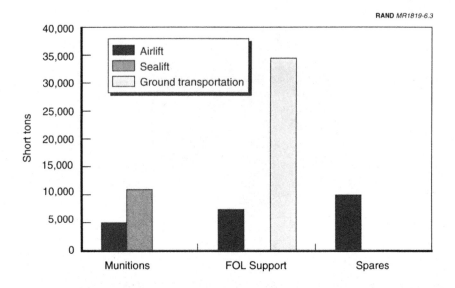

Figure 6.3—Modes of Transportation Used to Move Combat Support
Materiel During OEF

Fuel. During OEF, some fuel was trucked into such sites as Al Udeid from the refinery, some was flown in on C-130s to such sites as Jacobabad in support of the Special Operations Forces, and some fuel was moved by sea into such sites as Diego Garcia. Other sites had direct delivery from a pipeline. All modes of transportation of fuel are included in the total fuel-movement requirement shown in Figure 6.1. The total fuel moved during the first 100 days of OEF was approximately 800,000 short tons.

Al Udeid provides a specific example of fuel movement. On average, it received 55 truckloads of fuel each day during the first 100 days of the operation. However, on over 10 percent of those 100 days, 80+ truckloads of fuel were delivered to Al Udeid. On the heaviest day, 92 truckloads were delivered. Fuel, by far, dominated the commodity movements.

Munitions. Combat aircraft expended over 7,000 tons of munitions during OEF, a good portion of which was precision-guided. As with fuels, munitions were moved by various modes of transportation. The total munitions movements were approximately 16,000 tons, including approximately 5,000 tons airlifted in the form of standard air munitions packages (STAMPs), 3,000 tons moved by sea from CONUS, 8,000 tons used from aboard the Afloat Prepositioning Fleet (APF), and a few hundred tons as part of the bomber deployments themselves.

During OEF, the British protectorate of Diego Garcia served as the primary Air Force bomber FOL. Diego Garcia is one of four FOLs around the world identified for use by Air Force heavy bombers in the event of crisis. These so-called bomber islands are places where significant infrastructure and reserves of materiel can be built up ahead of time. JTF NA's experience had highlighted the importance of such bomber islands. However, in 2001, Diego Garcia was still under development. The Air Force was in the process of stocking munitions at Diego Garcia. Additional munitions were brought into Diego Garcia from CONUS during the buildup phase, just before the start of bombing operations. These included Joint Direct Attack Munition (JDAM) kits and Wind Corrected Munitions Dispensers (WCMDs), as well as the heavy bomb bodies themselves. But OEF operations began before the stocking was complete (HQ USAF, 2001a).

An analysis of all the cargo airlifted to Diego Garcia shows that much of the tonnage, approximately 54 percent, was devoted to movement of munitions. Munitions from the STAMP inventory were delivered into the theater by AMC military and chartered aircraft in a matter of days. However, it took approximately 60 missions to deliver approximately 3,800 tons of munitions, a substantial strain on strategic lift resources. Of the munitions tonnage total, the vast majority, approximately 28 percent of the total lift, was bomb bodies.[10]

That bombs accounted for a high percentage of the tonnage moved is perhaps not surprising; after all, bombs are very heavy. That so much airlift was devoted to carrying dumb bombs, however, is of particular interest. Valuable airlift could be saved if the bomb bodies, which are relatively inexpensive, plentiful, and heavy, were prepositioned at likely bomber forward operating locations or aboard the Afloat Prepositioning Fleet. Airlift could then be used to

_____

[10]Data were abstracted from the OEF TPFDD.

quickly move the more-expensive, scarce, and lighter components, such as guidance kits, on an as-needed basis.

The Air Force maintains inventories aboard the three ships of the Air Force APF. Normally, these ships are deployed forward in different regions of the world. However, at the time of Operation Enduring Freedom, one of the ships, the MV *Buffalo Soldier*, was being off-loaded at the Military Ocean Terminal, Sunny Point, North Carolina. Its cargo, normally stored in bulk format, was being transferred to containers for storage aboard the MV *A1C William H. Pitsenbarger*. As a result, the Air Force had only two of its APF ships deployed forward, which led to some reluctance on the part of the Air Force to release APF assets for OEF. In addition to using APF assets from the MV *MAJ Bernard F. Fisher*, the Air Force contracted the sealift vessel *Cornhusker State* to bring assets from the *Buffalo Soldier*, as well as other assets destined for the *Pitsenbarger*, to Diego Garcia. Large quantities of munitions were delivered by sealift. However, it took approximately 28 days for the *Cornhusker State* to sail from the East Coast to Diego Garcia.[11] Although it took longer to have the munitions moved by sea than by air, with enough time, munitions were in place when needed.

FOL Support Assets. Whereas munitions were moved primarily by air and sea, FOL support assets were moved mainly by ground transportation during OEF (see Figure 6.3). Most of the FOL support came from FSLs in the AOR; however, some was transported by air from CONUS.

Delivery of FOL support assets was sometimes faster for equipment coming from CONUS than for equipment coming from within the AOR. While deliveries to FOLs from FSLs in the same country were quick, an average of approximately four days, FOL support transportation time to FOLs from FSLs in another country could be much slower, ranging anywhere from two to four weeks. In contrast, FOL support deliveries originating from CONUS closed in four to 15 days.[12] Air transportation from CONUS and that from FSLs in the same country were the quickest methods of FOL support deliveries.

---

[11]Data were provided by ACP.

[12]Abstracted from data provided by CENTAF.

Many reasons can be given for the slow FSL intercountry delivery times—for example, limited TDS capacity, WRM warehouse throughput constraints, or reception host-nation agreements. In one case, the receiving base did not have the personnel to construct FOL assets, so the base requested that assets be held at the FSL. Intratheater airlift, especially in the early days of OEF, was in extremely short supply, with only a few C-130s in-theater. The lack of cargo aircraft was due not to a lack of airlifters in the fleet but to a lack of beddown space at the various FSLs, which were also serving as combat, ISR, and tanker bases.

Trucks were contracted locally but were subject to availability, road conditions, and, for some sites, the availability of ferries. In some cases, Air Force spare parts piggybacked on trucks that were contracted by the WRM contractor (DynCorp) to carry FOL support equipment. Locally contracted trucks presented a force-protection concern, requiring additional inspections, escorts, or transloading, all of which required additional time.

Spares. Spare parts, which accounted for only 1 percent of the total sustainment movement during OEF,[13] are a small but vital part of sustainment movement. OEF success depended on the availability of the Air Force's HD/LD assets. The small fleet sizes of ISR assets (U-2, Predator, Global Hawk, and E-3) and AFSOC fixed- and rotary-wing aircraft demanded immediate spare parts support from CONUS. Movement of spares is dependent upon the theater distribution network, especially movement by air.

In the movement of spare parts, no single source of air transportation was best for every destination. Transportation-time data during OEF show that commercial carriers were faster at some locations and that AMC was faster at others. Furthermore, the performance of AMC relative to commercial carriers, and even that among the different commercial carriers, varied from week to week. To illustrate this point, Figure 6.4 displays some spares transportation data collected

---

[13]Data were provided by AFMC/LSO-LOT.

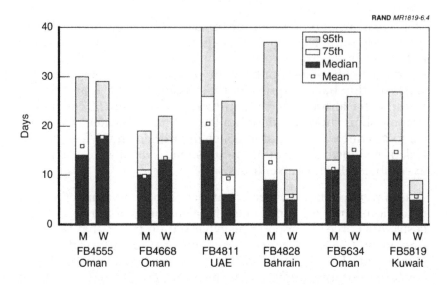

NOTES: In these data, airlift times begin when the cargo leaves the CONUS
Defense Logistics Agency (DLA) distribution center and ends when the cargo receipt
paperwork has been completed by the Transportation Management Office or base
supply. M=military airlift; W=World Wide Express (WWX) airlift.

**Figure 6.4—For Air Transportation of Spares, No One System Is Best in All
Cases**

by the Air Force and the RAND SDMI project from October through
December 2001. The squares on the graph show the average military
airlift times (M) and average World Wide Express (WWX) airlift times
(W) to several locations in Southwest Asia in support of OEF.[14] Also
shown on the figure are the median, the 75th percentile, and the 95th
percentile.

Management of Theater Distribution

Since no one mode of transportation best serves the needs of all locations, those charged with ensuring prompt resupply of materiel need
to select the service that best meets their needs. As experienced in

------

[14]Data were provided by SDMI.

JTF NA, multiple supply chains need to be used to ensure responsive delivery to warfighters at different locations. To make proper use of these different supply chains, planners must have ready access to information that enables them to make good transportation decisions.

Information feedback and decisionmaking loops could be improved. There were coordination problems and gaps between the TDS and the strategic movements system—the intertheater movements system—during JTF NA and then again during OEF. TDS was slow to evolve, and intertheater and intratheater movements were not well coordinated. Many problems arose in establishing a theater distribution system to meet Air Force needs. They began with the Air Force playing a larger role in the development and design of the TDS than expected.

TDS Responsibility and Organization. According to doctrine, the combatant commander designates which service will have responsibility for the Joint Movement Center and TDS—for the planning and execution of all movements of materiel and personnel within the AOR by land (trucks and rail), sea (ships and barges), and air. For this responsibility, the combatant commander may designate the service that is most capable of performing the tasks or the predominant user of the system (Joint Chiefs of Staff, 1996). In past operations, development of the TDS in the Central Command AOR was an Army responsibility. However, at the beginning of JTF NA the air component had the preponderance of forces, so the Air Force was given responsibility for TDS. The same situation occurred during OEF. At the start of OEF, the Army had few forces in theater and since resupply to geographically dispersed, austere areas was largely by air, Central Command delegated responsibility for the JMC and TDS to the Air Force through the AFFOR A-4 (which also acted as the Combined Forces Air Component Command C4).

The TDS is vitally important for meeting the rapid deployment and resupply needs of the AEF. In JTF NA, as in OEF, many problems arose in establishing a TDS to meet Air Force needs. In JTF NA, these problems began with the Air Force playing a larger role in the development and design of the TDS than had been anticipated. Air Force personnel assigned this responsibility may not have the training or background necessary to develop a TDS, an area in which future

Logistics Readiness Officers (LROs) with proper training might be expected to serve.

The OEF theater distribution system evolved over time. The JMC was collocated with the Air Mobility Division (AMD) in the Combined Air Operations Center. This collocation did help in planning and coordinating air movements; however, the JMC and TDS involve all modes of transport, and placement should be carefully reviewed.

As OEF unfolded, CAOC personnel working TDS had a difficult time projecting distribution system requirements. This same problem existed in JTF NA. Unable to come up with good estimates for the TDS, the initial OEF TDS relied on the TDS transportation assets that were in place to support Operation Southern Watch: 4 C-130s—a capacity that proved inadequate to meet OEF TDS needs. A few months into OEF, large backlogs of cargo developed at transshipment points in the AOR. It was not until several months later that standard air routes (STARs) were established to meet OEF TDS needs. This backlog is not consistent with AEF strategies of deploying light and lean and relying on rapid resupply.

Theater distribution was further complicated by the fact that Operation Southern Watch was going on in the same theater at the same time. Prioritization between different ATOs and associated FOLs was difficult.

During OEF, hubs were established to receive inbound shipments, then to ensure those shipments' continued movement to their final destination in the AOR. The effectiveness of this process was hampered in the early days by an immature intratheater transportation system. STARs were not established during the first 100 days of the operation. Large backlogs of cargo developed at transshipment hubs in the AOR (HQ USAF, 2001b; Barthold, 2002, p. 5), peaking at 1,000 pallets and persisting during the first 100 days of the operations. The TDS transportion assets then consisted of 10 C-130s for intratheater lift—not enough to work down the persisting backlog. Finally, the Director of Mobility Forces requested, and USTRANSCOM approved, the transfer of tactical control of a number of C-17s, typically used in an intertheater role, to work down the backlogs. Without host-nation support to bed down the C-17s, the aircraft were moved from base to base for a few days, working down the backlog, and were then returned to USTRANSCOM.[15] Some (SOF) units sent their own air-craft to pick up critical parts at transshipment points.

ITV and Communications. In-transit visibility was also a problem during OEF. After Operation Desert Storm, transporters had to track ITV on only 4 C-130s in the Central Command AOR.[16] With more aircraft in the AOR in support of OEF, ITV became much more diffi-cult to track. ITV was often lost on units and on individual personnel once they left their home stations (Barthold, 2002, p. 4). Visibility was lost at transshipment points, such as Rhein Main, where large shipments were subdivided into smaller shipments going many places. Cargo from one C-5 did not always fit into the cargo area of two C-17s, which caused some shipments to be subdivided.[17]

Communications were also an issue. The bandwidth allocated among the mobility forces was small. Firewalls and configuration problems interfered with reliable ITV.[18] In addition, different bases had different communications infrastructures. Army bases differed from Navy bases, which differed from Air Force bases. Some bases used satellite communications; some used landlines. The combina-tion of transshipment problems and lack of reliable communications resulted in lost cargo for short periods of time.

Ramstein had ITV difficulties because of the sheer volume of materiel and personnel passing through the base. New software, the Deployable Global Air Transportation Execution System (DGATES), was put in place at the beginning of OEF. The Air Force accelerated the installation of DGATES to assist with ITV. Reportedly, data from the Global Air Transportation Execution System (GATES) were useful, but DGATES had some problems.[19]

Since DGATES was a new system, training could have been an issue. In addition, DGATES and other systems rely on human input. The personnel inputting the data need to understand the importance of

---

[15]Interview with Maj Gen Richard Mentemeyer, October 2002.

[16]Interview with Maj Gen Richard Mentemeyer, October 2002.

[17]Interview with Mr. Paul Galloway, AMC/XPD, March 2002.

[18]Interview with Lt Col Thomas Klincar and Mr. Paul Galloway, AMC/XPD, May 2002.

[19]Interview with Mr. Frank Weber, USTRANSCOM J-3/J-4, May 2002.

the data and the result of inaccurate/incomplete data. For example, during OEF, units would arrive at the aircraft without proper documentation. The units would be shipped even without the proper paperwork,[20] which could cause a loss of visibility for the shipment.

Currently, no office at AMC is responsible solely for ITV. An ITV cell, operating 24 hours a day, 7 days a week, was set up during OEF to track missions down to Level 4 data.[21] The Air Force Inspection Agency is currently examining ITV issues in an effort called Eagle Look.[22] In past exercises and contingencies, ITV has not been emphasized; more focus could be placed on ITV during future exercises.

Using Information to Close the Loop. TDS personnel struggled with early predictions of how much intratheater lift would be required, even after initial assessments had been made. Attention was given to monitoring how well the system was performing. For example, early decisions were made about the extent of backlog that would constitute "adequate performance," raising the questions, What constitutes good "end-to-end" distribution times? and What metrics are needed for TDS?

The end-to-end system needs to be able to measure distribution times against those needed to meet operational objectives. Chapter Three, which describes some of the CIRF CSC2 features, shows the feasibility of such a system.

Figure 6.5 shows what happens when the strategic and theater systems are not coordinated. It gives a breakdown, by hub, of the Air Force and Army shipments that sat awaiting transportation in support of OEF during December 2001 and January 2002 at two hubs in the AOR, one intertheater hub, and two CONUS hubs. Each bar on the graph represents shipments going to a separate location. For example, AOR Transshipment Hub 1 shows the hold times for

---

[20]Interview with Mr. Frank Weber, USTRANSCOM J-3/J-4, May 2002.

[21]Interview with Mr. Frank Weber, USTRANSCOM J-3/J-4, May 2002.

[22]Interview with Lt Col Cornell and Ms. Lori Jones, USTRANSCOM ITV Working Group, May 2002.

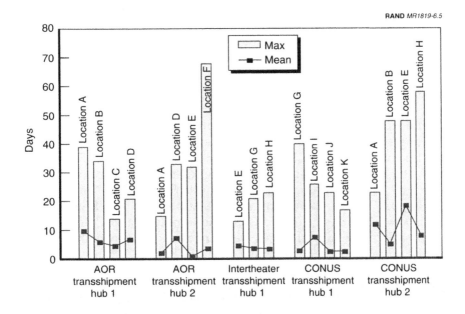

Figure 6.5—Materiel Sat at Transshipment Hubs During OEF Because the Strategic and Theater Systems Were Not Well Coordinated

shipments going to four different locations—A, B, C, and D—from the one hub. Because of the backlog issue, materiel spent a significant portion of time sitting at these hubs. The time spent at the hubs includes the time that cargo waited for the TDS to deliver cargo from transshipment points after being downloaded by the strategic movements system.[23] The figure shows that these wait times did exist, and not only at hubs in the AOR. Shipments sat at CONUS ports as well as at strategic hubs, waiting to get into the AOR.

These wait times could be improved through better coordination of the theater distribution system and the strategic movements system. Other options for improving the total end-to-end distribution times could involve placing one agency—for example, USTRANSCOM—in charge of developing an end-to-end military distribution system.

---

[23]These hold times could be a function of combatant commander priorities and not necessarily how fast the movement could occur, given a higher priority. Data provided by SDMI.

Such an agency would operate similarly to commercial carriers—for example, Federal Express—that are responsible for end-to-end performance.

IMPLICATIONS

Joint doctrine indicates that TDS responsibility can be appointed to any service based on "either the dominant-user or the most-capable-service concept" (Joint Chiefs of Staff, 1996, p. v). During JTF NA and OEF, the combatant commander assigned theater distribution responsibilities to the Combined Forces Air Component Command. During OEF, the combatant commander stipulated that TDS responsibility would transfer to the Combined Forces Land Component Command once ground forces were engaged, but this had not occurred more than a year and a half after OEF began.

The assignment of TDS development to the Air Force in OEF and the large role that the Air Force played in developing the TDS in JTF NA beg the question of whether the Air Force will assume this responsibility in the future. It is likely that the air component will have the "predominant" portion of the forces in the early phases of future contingencies; therefore, the Air Force could be tasked with TDS in the future.

JTF NA demonstrated what transportation capabilities were achievable, given the robust transportation networks available in Western Europe. But, even with a well-developed transportation infrastructure, TDS performance varied and adaptations were made to mitigate shortfalls during JFT NA. As in OEF, not all operations will be conducted where well-established transportation infrastructures are already in place.

The transportation system used during any operation will be complex and multimodal, and will involve numerous customers (for example, Army, coalition, and Air Force). If the Air Force is asked to be responsible for TDS or even if it just provides input to another service that controls TDS, the Air Force needs to provide education and training to effectively plan and manage TDS responsibilities. Staff must be equipped with training and information necessary to make informed decisions. Creation of a Logistics Readiness Officer shows promise to fulfill this critical need.

Nonetheless, a specific education and training plan for theater distribution needs to be developed. Theater distribution is more than just the onward movement of spare parts. The system also includes a network to link FSLs and CSLs to FOLs. MAJCOM components need to work with USTRANSCOM to develop integrated plans to make the transition from peacetime operations smoothly into wartime operations. An expeditionary Air Force cannot afford critical assets, to include materiel moving to forward operating locations and unserviceable items being returned to repair locations, to sit backlogged at FSLs and transshipment points.

Options for having a single party develop an end-to-end military system instead of a strategic movements system and a TDS need to be explored. The difference between a strategic movements system and a tactical movements system is not clear. For instance, is a system that connects CIRFs or supply FSLs located in one AOR to FOLs in another AOR, as happened in OEF, a strategic system or a TDS? If it is a TDS, which combatant commander should set up the inter-AOR system, the supporting commander or the supported commander? Perhaps the separation of the TDS and the strategic movements system has outlived its usefulness.

However, another solution may be to develop Distribution Units in each service that would be trained to fill in the gaps between the strategic and the tactical distribution systems. These units would be similar to a Federal Express or a United Postal Service regional office. They could have common training, tools, and performance metrics and could seamlessly merge into the TDS gap during contingency operations.[24]

No matter which solution is chosen, the system must be able to support the global War on Terrorism and the global positioning of combat support resources to meet commitments across a wide variety of scenarios.

---

[24]For more information about this Distribution Unit concept, see Halliday and Moore (1994).

Chapter Seven

RESOURCING TO MEET CONTINGENCY, ROTATIONAL, AND MRC REQUIREMENTS

Combat support resources are allocated and employed in meeting today's AEF rotational and contingency requirements in ways that are not consistent with the assumptions that are made in the current resource requirements determination processes. This chapter analyzes how resource planning factors and processes may need to be changed to better meet the needs of today's expeditionary air and space forces and current defense programming guidance.

FINDINGS

At the times of the JTF NA and OEF engagements, many global operations were being conducted. Those operations drew on combat support resources to a significant degree and affected residual combat support capabilities. Some of these other operations are

- Operation Southern Watch
- Operation Northern Watch
- Support in Bosnia and Kosovo
- Operation Infinite Reach
- Operation Determined Response
- Operation Noble Eagle
- Operation Border Support
- Antiterrorism support.

The following paragraphs give an overview of the personnel and re-sources used in other ongoing operations.

Operation Southern Watch and Operation Northern Watch over Iraq have been ongoing operations since the early 1990s. Operation Southern Watch involves more than 6,000 Air Force personnel sup-porting over 28,000 sorties. Over 1,400 coalition and Air Force per-sonnel and 45 aircraft are deployed in support of Operation Northern Watch. WRM is currently being used in support of both of these on-going operations. (In theory, WRM assets are intended to be saved for use during major regional conflicts.)

JTF NA involved over 44,000 airmen supporting over 30,000 missions, and OEF involved over 12,000 personnel supporting over 11,000 missions. Looking at other ongoing operations, we see that Operation Noble Eagle involved over 35,000 personnel supporting over 13,000 sorties (www.GlobalSecurity.org). In addition, Exercise Bright Star, which was being conducted in Egypt from October 8, 2001, through November 1, 2001, involved about 23,000 U.S. military. The Georgia Train and Equip program, which began in April 2002, involves 150 U.S. military personnel training forces of the former Soviet republic of Georgia in counterterrorism capabilities. It is ex-pected that this sort of program will be used in as many as 20 other countries in the future.

In addition to all the ongoing operations mentioned above, other smaller-scale operations have been conducted—for example, Operation Infinite Reach in Afghanistan and the Sudan in 1998 and Operation Determined Response in 2000.

All these ongoing operations have strained the capability of the com-bat support community to open and sustain forward operating loca-tions for other engagements. The ability to indicate what capabilities exist within an AEF deployment cycle[1] does not exist, and methods are needed to develop these capability estimates from available equipment and personnel resources.

---

[1] The AEF model evenly distributes Air Force capabilities into 10 parts, often referred to as "buckets." The buckets are paired into five deployment cycles. The *deployment cycle* is the time period in which the bucket is eligible to be deployed.

In this chapter, we consider the resourcing of the following:

• Harvest Falcon (HF) and FOL support assets

• Munitions

• Personnel.

Harvest Falcon and Other FOL Support Assets

Table 7.1 illustrates how planning factors, which are used to determine Harvest Falcon (HF) requirements, differ significantly from how HF assets are employed today. The Harvest Falcon planning factors are shown on the left side on the table; current employment factors are on the right. The planning factors are based on supporting full-size squadron deployments to a bare base with adequate room to set up Housekeeping, Flight Line, and Industrial Operations sets. But, JTF NA and OEF experiences show that

Table 7.1

Harvest Falcon Planning Factors Versus Actual Usage Today

Harvest Falcon Resource Planning Factors	Harvest Falcon Current Employment Factors
Bare base deployment; space and latitude to build to economies of scale.	Deployment to existing bases to augment infrastructure. Must fit in space available.
Short, intense wartime involvement; minimal infrastructure to generate sorties.	Sustained, indefinite deployment/employments; additional quality-of-life and force-protection requirements.
MRC full-squadron deployments.	Less-than-squadron deployments, and modular FOL support.
High-threat force-protection requirements not included.	Significant additional requirements for FOL support modules/items.
Support to Air Force units only.	Support for other services .
Harvest Falcon requirements, FSL, and distribution throughput are computed against specific planning scenarios.	Harvest Falcon sets have been used to support other AORs routinely—e.g., support of Burgas in OEF, other USAFE sites in JTF NA, and throughput needs to be computed to meet global AEF goals.

numerous Air Force deployments involve less-than-squadron-size units deploying to coalition-partner military sites. The deploying forces may use existing infrastructure but require additional assets—for example, power-distribution units; and, because of space limitations, detached facilities may have to be built in a restricted amount of space. Further, specific components of sets are issued to meet specific demands for force protection or other needs—for example, light sets. Requirements planning factors also assume that the sets would be used one time to meet MRC needs; however, today, the sets are being used to sustain long-term permanent rotations, such as Operation Southern Watch.

The last row in Table 7.1 addresses an important discrepancy between planning factors and AEF support requirements: Specific planning scenarios are used to determine requirements, whereas the required assets are those needed in the reality of meeting AEF support requirements worldwide. FSL and distribution throughput needs to be determined to meet global AEF deployment goals in addition to specific theater needs. It may be that the global goals are more stringent than specific theater needs.

Figure 7.1 shows how the difference in planning assumptions and employment factors has created widespread shortages in particular Harvest Falcon components and in mission-capable and -deployable sets during OEF.

Primarily, Figure 7.1 shows the specific high-demand components of Harvest Falcon sets that are issued to support deployments and are removed from complete sets to meet demands. In all but the shower/shave units, demand and operational needs during OEF exceeded the planning-factor authorizations.

As shown, over 150 lighting units and expandable common-use shelters have been issued to support OEF operations. For the MEP-12 generators, those above-the-authorization demands effectively eliminate the high-voltage power for an additional 12 Housekeeping sets.

Therefore, specific commodities, not complete sets, are issued to meet needs; and commodities, not complete sets, tend to remain issued rather than being returned to storage, and they tend to remain

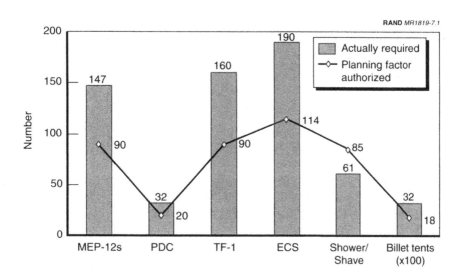

SOURCE: Data are from CENTAF/A-4-LGX.

NOTES: MEP-12=power generators; PDC=power-distribution center; TF1=lighting units; ECS=expandable common-use shelters.

Figure 7.1—OEF HF Employment Practices Differ from Planning Practices

in use for extended periods of time—differences between programming assumptions and current usage. These practices render complete sets incomplete and not ready for deployment.

Operations Northern Watch and Southern Watch have also constrained the availability of Harvest Falcon kits.[2] Beddowns for Operation Northern Watch/Operation Southern Watch were using Harvest Falcon kits before OEF even began. WRM assets that, in theory, are to be saved for use during major regional conflicts, in reality are being used for most operations in the AOR.

Because Harvest Falcon equipment and subcomponent unit type codes (UTCs) represent actual capabilities, CENTAF focuses management attention down to the equipment level to paint a truer picture of the health of the theater capability. For example, the most

---

[2]Interview with Maj Dennis Long, CENTAF/A-4 LGX, September 2002.

frequently used items for a Housekeeping set are the MEP-12s and the power-distribution center (PDC).[3] *If the Air Force could purchase more power-generation items or target those specifically for reconstitution, then more nonusable sets at the prepositioning site could be brought back to usable condition.*

Management by specific subcomponent UTCs (that is, commodities rather than sets) not only paints a truer picture of current status, it also allows Air Force planners to see high-use items and can facilitate the lessening of the effects of their shortfalls on the entire system. *Again, resource planning factors and management techniques need to come into line with current AEF usage patterns.*

Munitions

Table 7.2 shows some of the major differences between the resource planning factors for munitions and the actual usage of munitions in current contingency actions, including JTF NA and OEF.

The emphasis on reduced collateral damage and the efficiency in identifying targets have placed a premium on the use of precision-guided munitions (PGMs), such as laser-guided bombs and the Global Positioning System (GPS)-guided JDAM, in recent operations. As shown in Figure 7.2, the use of precision weapons increased almost twofold in OEF from that during JTF NA.[4] JDAMs were especially heavily used because of the accuracy of GPS in any weather condition, although not as accurate as laser-guided bombs. Although designed to be used against traditional high-value fixed targets, such as command and control nodes, they were heavily used against caves and against enemy ground troops in close air support missions flown by bombers at relatively high altitudes.

These "smart" munitions are expensive and limited in supply when compared with "dumb" bombs. Both the Air Force and the Navy used precision-guided munitions during OEF, reducing the available

---

[3]Data are from CENTAF/A-4-LGX

[4]JTF NA OEF data were abstracted from TPFDD.

Table 7.2

Munitions Employment Versus Planning Factors

Munitions Resource Planning Factors	Munitions Current Employment Practices
Specific scenarios, aircraft types, and target sets are used to compute requirements.	Actual scenarios differ from planning scenarios.
Precision munitions are determined against specific targets in the scenarios.	Precision munitions are the munitions of choice, owing to tight rules of engagement on collateral damage.
Computations assume that munitions will be used in specific scenarios and are distributed to specific combatant commanders for anticipated use in specific AORs.	Munitions that are distributed to specific AORs are used in other AORs routinely.
Munitions FSL throughput and distribution requirements are determined, if completed, based on specific scenario considerations—e.g., trucking capacity in Korea.	Munitions FSL throughput and distribution capabilities need to be based on *global needs*.

RAND *MR1819-7.2*

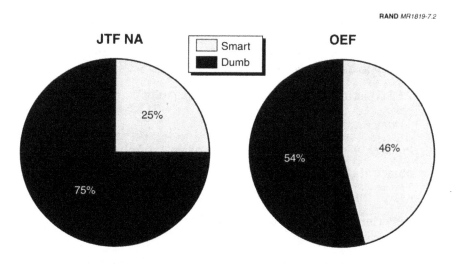

Figure 7.2—Use of PGMs During JTF NA and OEF

stockpile. The Air Force went to great lengths to ensure that adequate munitions were available when needed. Bomb bodies, guidance systems, fin kits, and fuzes were both airlifted and sealifted to the point of need. The Air Force had to "borrow" PGMs from other theaters because of the limited inventory of these assets. This borrowing of assets could have hampered the Air Force's ability to engage in another contingency had one occurred in another theater immediately on the heels of OEF.

Personnel Issues

Disconnects between resource planning assumptions and actual AEF employment factors can create not only resource shortages, requiring the Air Force to make a concerted effort to reconcile planning and employment factors, but also personnel issues. The Air Force AEF model was designed to evenly distribute resources into 10 parts, often referred to as "buckets." Each bucket has roughly equal capability. The AEF deployment buckets are paired into five deployment cycles, when a bucket would be eligible for deployment. The design of the combat support resources portion of the AEF model was intended for each base to be tasked to provide assets only twice in a 15-month cycle, both combat support and aircraft. The intent of the AEF construct was for deployments to be held to a minimum and for personnel to know when and for how long they were going to be deployed.

The AEF construct consists of the following "rules":

- 90-day rotations

- no base to ask for assets more than twice per AEF cycle

- library of UTCs for each AEF

- every airman in a UTC

- AEF Center suggests potential candidates for ACS UTC requirements to units.

However, these AEF "rules" were violated during OEF for several combat support fields, including force protection, communications, civil engineering, and fuels. As a result, the AEF Center had to reach forward and deploy personnel scheduled in future AEF deployment cycles. Figure 7.3 shows the current forward-reaching and extended tours for personnel as of November 2002 (Przybyslawski, 2002, chart 10). The Army has also experienced higher personnel tempo in recent years.[5]

To illustrate the personnel shortages, we examine three overextended career fields: security forces, fuels, and communications.

Security Forces. The security forces Air Force Specialty Code (AFSC) has been extremely overextended since September 11, 2001. As of January 13, 2002, approximately 35 percent of all personnel deployed in support of OEF were security forces personnel. Looking only at the security forces AFSC, approximately 30 percent of all security

RAND *MR1819-7.3*

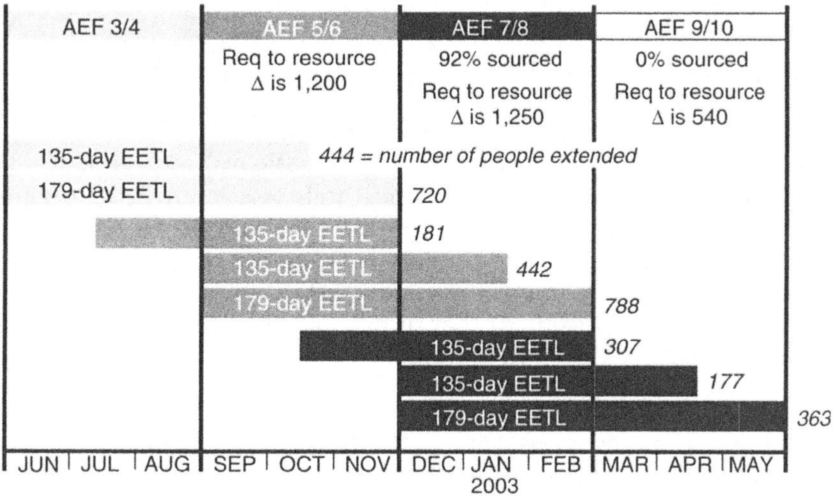

AEF 3/4	AEF 5/6	AEF 7/8	AEF 9/10
	Req to resource Δ is 1,200	92% sourced	0% sourced
		Req to resource Δ is 1,250	Req to resource Δ is 540
135-day EETL	444 = number of people extended		
179-day EETL	720		
135-day EETL	181		
135-day EETL	442		
179-day EETL	788		
135-day EETL	307		
135-day EETL	177		
179-day EETL	363		

JUN JUL AUG SEP OCT NOV DEC JAN FEB MAR APR MAY
2003

SOURCE: Przybyslawski, 2002, chart 10.
NOTES: Data as of November 11, 2002. EETL=Estimated Extended Tour Length.

Figure 7.3—AEF Rotational Cycle Extensions as of November 2002

---

[5]For more information about Army personnel tempo issues, see Polich et al. (2000) and Sortor and Polich (2001).

force personnel were deployed. This 30 percent does not include the additional security personnel assigned to their own home base to support Operation Noble Eagle or to support additional security requirements in CONUS resulting from the terrorist attacks of September 11, 2001 (USAF, Air Force Operations Group, 2002).

In addition to active-duty personnel, the Air National Guard (ANG) activated a large number of security forces. ANG security forces were used at home bases, overseas, and as airport security support—a job for which they are not trained. Approximately 10 percent of the ANG security forces personnel deployed overseas in support of OEF. Again, this number does not include personnel assigned to their own home base to meet higher security requirements associated with the post-9/11 world.

Fuels Personnel. The fuels community was also hit especially hard during the early days of OEF. Operational requirements increased at AMC hubs as deployment airlift increased. For example, in the month of October, Ramstein AB, Germany, had 5 percent of its fuels personnel deployed and experienced a 103-percent increase in the amount of fuel issued over the same time the previous year. Tanker bases required more fuel to support fighter aircraft flying Civil Air Patrol (CAP) over the nation's cities, and requirements increased at other bases as squadrons prepared to deploy. Another example is MacDill AFB, Florida. In November 2001, 27 percent of its fuels personnel were deployed. At that time, MacDill experienced a 4.5-percent increase in the amount of fuel issued.

The decision to reach forward into future AEF deployment cycles was not taken lightly. However, under the current method of sourcing personnel, the current rotation schedule, and contingency requirements, it had to be done. Figure 7.4 shows the number of fuels personnel that are currently authorized and funded. Although it appears that a large number of fuels personnel have been authorized, note that contractors and civilians are not deployable. In addition, Reserve personnel may only deploy for 15-day rotations, unless they are called up for duty.

Figure 7.4 also shows the current AEF deployment construct for fuels personnel, the requirements for fuels personnel for OEF, and the cur-

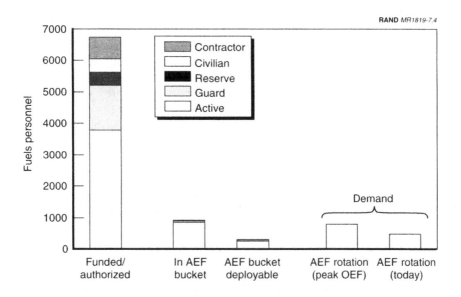

Figure 7.4—Demand for Fuels Personnel Was Double What Was Available in the AEF Bucket

rent AEF rotations. As shown on the right side of the figure, the OEF deployment requirement for fuels personnel was about double the current rotational requirement. The figure also shows that the number of deployable fuels personnel assigned to a given AEF bucket is not large enough to cover current AEF rotations.

Upon review of fuels-unit AEF requirements, the fuels community found that it could deploy only 20 percent of the personnel from each location because of home-station requirements. When it looked at the workload decrease expected to accompany a deployment, it found the decrease to be approximately only 4 or 5 percent. Using the current AEF construct, in which each base is tasked twice to provide fuels personnel, we see that the fuels community was tasked for 40 percent of its available force during each AEF cycle, stressing the home base operations.

Combat Communications Personnel. The requirement for combat communications personnel placed a large burden on that career field. While there was a requirement to stand up new bases in an austere environment and establish communication links for such new systems as Predator, competing requirements were keeping the communications community busy inside CONUS. CONUS forces were trying to turn outward-looking radars inward to establish communication linkups between North American Air Defense Command (NORAD), the White House, and CONUS bases, where armed fighters were postured on alert status in support of Operation Noble Eagle.

A System for Assessing Capability

The AEF Center and others routinely perform resource assessments like the one for the fuels community outlined for combat support skills; however, such assessments do not provide insights into some very fundamental questions, including (Hornburg, 2002):

1. What combat support capabilities exist in the AEF to open bases after steady-state rotational commitments have been met?

2. What combat support capabilities exist in the AEF to augment capabilities at preexisting yet not fully developed bases?

3. What are the limiting constraints on capability, equipment, or personnel?

4. What are options for mitigating the constraints and what are their costs?

Senior defense planners and senior Air Force officers have laid out the requirement to provide this assessment capability, and RAND Project AIR FORCE has a project under way to develop such a capability. Appendix C highlights how this capability assessment could be conducted.

IMPLICATIONS

JTF NA and OEF findings indicate that current resource-planning factors and methods are not aligned with current resource-consumption factors. Combat support resources are stretched thin in meeting current rotational, peacekeeping, and training requirements and may leave little capability for meeting future small-scale contingencies (SSCs) or potential MRCs. We show that small-scale contingencies such as JTF NA and OEF may not necessarily require fewer support resources than an MRC. In fact, actual resource-usage patterns differ from those used in MRC planning computations; in some cases, SSCs may actually require as many resources—or even more.

One possible answer to the problems of limited resources and planning factors' not matching actual resource use would be to change the factors and increase the inventory levels of materiel, and to add personnel. Computations could be made to determine requirements as a function of the current combat support posture and policies. However, with many competing needs, the Air Force may not be able to afford this approach. Still, several options and trade spaces are available between alternative requirements, alternative combat support distribution options, and other support policies; they may be able to satisfy operational requirements more effectively than just increasing the size of existing pipelines, assuming the current way of providing combat support is the best way.

One such option would be to make investments to decrease delivery time—for example, by positioning items closer to the point of need, perhaps by distributing existing resources to more FSLs in differing AORs. Another option for decreasing delivery time would be to improve throughput capability of existing FSLs and associated distribution capability—for example, by increasing the working maximum-on-ground (MOG) at FSL sites or nearby airports, or by improving rail or sea handling capabilities. Additional ships to store and move WRM may improve delivery times to FOLs. Smaller, faster ships loaded with high-demand assets may help to alleviate some initial airflow concerns. An integrated analysis of options is needed.

Planning factors for determining WRM requirements and capabilities for global WRM distribution need to be considered jointly. Alternatives to stockpiling munitions and other WRM assets need to be considered in today's uncertain world. One approach may include constructing/reengineering flexible munitions production lines with surge capacity, beyond having on hand stocks needed to support the initial phases of possible contingencies.

Evaluating combat support options in today's uncertain world requires a capability-based assessment method to provide insights into the capabilities that exist to meet a wide variety of scenarios with alternative levels of investments in combat support resources. A "capabilities" view of resources—in which various investments would be stated in terms of what they could support—may be a more appropriate way to consider resource investments today. For example, one investment would be the ability to support X permanent rotations, a small-scale contingency of Y size (so many beddown sites), and an MRC of Z size (so many beddown sites). The Air Force does not know and will never know with certainty what scenarios it may be expected to support in the future, but it should have the ability to state what capabilities it can support from a combat-support perspective. Embryonic tools have been developed within RAND Project AIR FORCE in studies that have begun to establish the capabilities-based planning and assessment called for in defense guidance and by senior Air Force leaders.

Finally, the AEF is a transformational construct and has many implications for how resources will be provided and what types will be needed in the future. The major theme of substituting speed of deployment and employment for presence has significant resource implications. It also has significant implications for the types of resources that need to be procured. Getting the deployment of light and lean initial support packages quickly to the fight places emphasis on having reliable transportation and CSC2. The Air Force needs adaptive combat support.

Current AEF scheduling rules may be an effective and efficient manner for scheduling and deploying aircraft and aircraft support units; however, they may not be the best for scheduling ACS. Specifically, balances must be struck between disruption of home-station support and deployment commitments. Many options and alternatives exist for ACS scheduling rules. Those options should be evaluated with respect to how they affect the performance of home-station and deployed combat support.

Chapter Eight

CONCLUSIONS

OVERALL EVALUATION OF FIVE AREAS OF COMBAT SUPPORT

Table 8.1 summarizes our evaluation of combat support in the five areas investigated during this study.

Processes for combat support execution planning and control and organizational alignments have improved since JTF NA, but OEF

Table 8.1

Assessing Combat Service Support

	Operation Allied Force–JTF NA	Operation Enduring Freedom
Combat support execution planning and control	Ad hoc	Improved, but still ad hoc
Forward operating location development and site preparation	Varied	Varied
Forward support location and CONUS support location preparation and operation	Inefficiently used	Better linked to warfighter needs
Reliable transportation to meet forward operating location needs	Not prepared for responsibility	Inadequate; built on existing Operation Southern Watch system
Resourcing to meet contingency, rotational, and MRC requirements	Differed from planning factors	Differed from planning factors

demonstrates that more attention is still needed in this area. CSC2 was not well understood; consequently, the ad hoc organizational structure that developed varied from doctrine and continued to evolve throughout the operation.

Austere FOLs and an immature theater infrastructure during OEF emphasized the importance of early planning, knowledge of the theater, and FOL preparation. Even with a more-developed infrastructure, FOL developed during JTF NA was delayed by host-nation support and site surveys. Site surveys were ad hoc and nonstandardized in both JTF NA and OEF. Host-nation support was difficult to negotiate. The resulting deployment timelines varied widely in both operations.

The current AEF force structure of light, lean, and lethal response forces is highly dependent upon FSL capacities and throughput. Austere FOLs and the immature theater infrastructure illustrated the importance of using FSLs efficiently during OEF. Because of problems identified during JTF NA, improvements have been made in linking FSLs and CSLs to dynamic warfighter needs, but much more can be done in this area.

Realizing AEF operational goals depends on the presence of assured and reliable end-to-end deployment and distribution capabilities; therefore, these capabilities need to be configured quickly to connect the selected sets of FOLs, FSLs, and CSLs in contingency operations. Under current joint doctrine, the service with the preponderance of force may be delegated the responsibility for developing and operating the theater distribution system. Since the Air Force may be the predominant user of the TDS in early phases of future campaigns, the Air Force may be delegated the TDS responsibility.

Even if another service is delegated this responsibility, the Air Force should play an active role in determining TDS capacities and capabilities. AEF success depends on the early establishment of reliable and responsive TDS capabilities. The Air Force, as well as other services, depends on joint, global, multimodal, end-to-end transportation capabilities.

In both JTF NA and OEF, problems encountered with establishing a responsive TDS and those associated with integrating the strategic movements system with the TDS led to gaps in an end-to-end military deployment and resupply system that were not encountered by commercial carriers. During OEF, Federal Express and other carriers had end-to-end visibility and could track their responsiveness in meeting deliveries. This same kind of capability was not established until several months after operations began in the military portion of the transportation system.

Shortages in combat support assets, particularly in high-demand, low-density areas, such as combat communications, civil engineering, and force protection, overextended the AEF construct, resulting in the Air Force's borrowing against future AEFs during OEF. In addition, current AEF employment practices differ significantly from planning factors used in the Program Objective Memorandum (POM) process to provide for combat support resources.

To evaluate combat support options in today's uncertain world requires a capabilities-based assessment method. Such a method provides insights into the resources and skills that exist to meet a wide variety of scenarios with alternative levels of investments in combat support resources. A "capabilities" view of resources may be a more appropriate way to consider resource investments today than a scenario-based view.

Finally, the AEF is a transformational construct and has many implications for what types of resources and how resources will be provided in the future. The major theme of substituting speed of deployment and employment for presence has significant resource implications. It also has significant implications for the types of resources that need to be procured. Having to deploy light and lean initial support packages quickly to the fight places emphasis on reliable transportation and CSC2.

Current AEF scheduling rules may be an effective and efficient manner for scheduling and deploying aircraft and aircraft support units; however, they may not be the best for scheduling ACS. Specifically, balances must be struck between disruption of home-station support and deployment commitments.

RECOMMENDATIONS

Below is a list of the recommendations derived from the work on this study. These recommendations are suggested methods to improve Agile Combat Support for the AEF.

Combat Support Execution Planning and Control

- Establish clear doctrine for combat support execution planning and control.

- Clearly define command relationships.

- Integrate combat support planning with the operational campaign planning process.

- Develop control mechanisms.

FOL and Site Preparation

- Focus attention on political agreements and engagement policies.

- Standardize site-survey procedures and processes within the Air Force, with other services, and with U.S. allies.

FSL/CSL Preparation for Meeting Uncertain FOL Requirements

- Further develop the existing global network of FSLs and CSLs.

- Continue improvements in linking FSLs and CSLs to dynamic warfighter needs.

Reliable Transportation to Meet FOL Needs (TDS)

- Be prepared to play an active role in determining TDS capacities and capabilities

  — Identify lift requirements, including airlift, sealift, and movement by land

- — Initiate training and enhance personnel development policies to prepare for TDS responsibility
- — Work with joint commands to develop and resource plans to support the AEF with adequate TDS capabilities.
- • Review joint doctrine on the transportation system
  - — Consider having USTRANSCOM develop an end-to-end distribution system
  - — Consider establishing Distribution Units in each service to fill in TDS gaps during contingency operations
  - — Consider ways to improve TDS performance, including better in-transit visibility and demand-forecasting mechanisms.

Resourcing to Meet Contingency, Rotational, and MRC Requirements

- • Reevaluate current processes and policies for AEF assignments and the current POM assumptions with respect to combat support resources
  - — Align current employment practices with resource-planning factors.
- • Enhance the capabilities-based planning and assessment methods that RAND is currently developing.
- • Evaluate existing scheduling rules for combat support with respect to how that support will affect the performance of home-station and deployed combat support.

NODES AND RESPONSIBILITIES OF COMBAT SUPPORT EXECUTION PLANNING AND CONTROL (CSC2) TO-BE OPERATIONAL ARCHITECTURE

As we found in reviewing the lessons learned during operations in Serbia, CSC2 processes are well documented in neither current Air Force doctrine nor joint doctrine. As a result, understanding of the CSC2 process is limited in both the operational and combat support communities. This lack of understanding and an ad hoc organization resulted in problems in combat support command and control (C2) in both Joint Task Force Noble Anvil (JTF NA) and Operation Enduring Freedom (OEF).

In response to the CSC2 issues discovered during operations in Serbia, AF/IL asked RAND Project AIR FORCE to study the current CSC2 operational architecture and develop a TO-BE (future) CSC2 operational architecture (Leftwich et al., 2002). Over the course of two years, RAND Project AIR FORCE documented the current processes, identified areas in need of change, and developed processes for a well-defined, closed-loop[1] TO-BE CSC2 operational architecture that incorporated the lessons learned during JTF NA.

More specifically, the TO-BE CSC2 operational architecture (Leftwich et al., 2002) identifies the future CSC2 functions as including the ability to

---

[1]A *closed-loop process* takes the output and uses it as an input for the next iteration of the process.

- enable the combat support community to quickly estimate combat support requirements for force-package options needed to achieve desired operational effects and to quickly assess the feasibility of operational and support plans

- quickly determine beddown capabilities, facilitate rapid development of time-phased force and deployment data, and configure a distribution network to meet employment timelines and resupply needs

- facilitate execution resupply planning and performance monitoring

- determine the consequences of allocating scarce resources to various combatant commanders

- indicate when the performance of combat support deviates from the desired state and implement re-planning and/or get-well planning analysis.

This appendix presents the TO-BE CSC2 nodal[2] responsibilities and processes outlined in the CSC2 TO-BE operational architecture.

THE TO-BE NODES AND RESPONSIBILITIES

In the TO-BE architecture, a CSC2 nodal template is established, with clearly defined responsibilities for each CSC2 node. Table A.1 shows some of the important CSC2 nodes and their associated roles and responsibilities.

A key element of the TO-BE CSC2 operational architecture, this nodal template can ease the transition from a peacetime structure to a wartime structure. Specific organizations are designated to fulfill the responsibilities of each one of the nodes; however, the template allows for variations in organizational assignments by theater, and may even serve as a guide for configuring the C2 infrastructure while retaining standard responsibilities. Along with the template, having standing CSC2 nodes that operate in both peacetime and wartime

---

[2]A *node* is a point of intersection, within a larger infrastructure, where the integration of processes and information occurs.

Table A.1

TO-BE CSC2 Nodes and Responsibilities

Combat Support C2 Nodes	Roles/Responsibility
	Joint Staff
Logistics Readiness Center	Supply/demand arbitration across combatant commanders
	Combatant Commander
Combatant Commander Logistics Readiness Center	Combatant commander logistics guidance and Course of Action analysis
Joint Movement Center	Combatant commander transportation supply/demand arbitration
Joint Petroleum Office (JPO)	Combatant commander POL supply/demand arbitration
Joint Facilities Utilization Board	Combatant commander facilities/real estate supply/demand arbitration
Joint Materiel Priorities & Allocation Board	Combatant commander materiel supply/demand arbitration
	JTF
JTF J-4 & Logistics Readiness Center	JTF logistics guidance Supply/demand arbitration within JTF, among service components
	JFACC
Joint Air Operations Center Combat Support Reps	JAOP/MAAP/ATO production support
JFACC Staff Logisticians	JFACC logistics guidance
	Air Force
Air Force Contingency Support Center (CSC)[a]	Monitor operations Represent Air Force combat support interest to Joint Staff Conduct/review assessments of integrated weapons systems and base operating support Arbitrate critical resource supply/demand shortages across AFFORs
	AFFOR
Air Operations Center (AOC) Combat Support Element	JAOP/MAAP/ATO production support
AFFOR A-4 Staff (forward)	Site surveys/beddown planning Liaison with AOC combat support element
AFFOR A-4 Staff (rear) at an Operations Support Center (OSC),[b] which supports AFFOR A-4 Staff forward	Mission/sortie capability assessments Beddown/infrastructure assessment ASETF force structure support requirements Supply/demand arbitration within ASETF among AEFs/bases Theater distribution requirements planning Force closure analysis Liaison with Air Mobility Division in AOC Liaison with theater USTRANSCOM node

Table A.1 (continued)

Combat Support C2 Nodes	Roles/Responsibility
Deployed Units	
Wing Operations Center (WOC)	Disseminate unit tasking
	Report unit status
Combat Support Center	Monitor and report performance and inventory status
Supporting Commands (Force and Sustainment Providers)	
Logistics Readiness Center/CSC	Monitor unit deployments
	Allocate resources to resolve deploying unit shortfalls
Deploying Units	
Wing Operations Center (WOC)	Report unit status
	Disseminate unit tasking
Deployment Control Center (DCC)	Plan and execute wing deployment
	Report status of deployment
Commodity Control Points (CCPs)[c]	
Munitions, Spares, POL, Bare Base Equipment, Rations, Medical Materiel, etc.	Monitor resource levels
	Perform depot/contractor capability assessments
	Works with the CSC to allocate resources IAW theater and global priorities
Sources of Supply (Depots, Commercial Suppliers, etc.)	
Command Centers	Monitor production performance and report capacity

[a]Some of these functions, which will be performed by the CSC, were referred to as Global Integration Center (GIC) functions in Leftwich et al. (2002). The Air Force will not use the GIC name in implementation efforts; rather, it will associate GIC functions with the CSC.

[b]The functions performed by the AFFOR A-4 forward and rear need not be the same for all theaters or regions. The idea is to codify the responsibilities by COMAFFOR in each region before contingencies begin. OSC A-4 will have virtual Regional Supply Squadron representation at the OSC. Many of the spares-related command and control functions would be conducted at the RSS with OSC A-4 input and coordination. The same is true for ammunition control points.

[c]The CCP was referred to as a Virtual Inventory Control Point (VICP) in Leftwich et al. (2002) and in several articles associated with the Spares Campaign and Depot Re-engineering and Transformation. CCP will replace the VICP name.

can ease the transition from daily operations to higher-intensity operations, allowing the Air Force to train the way it intends to fight.

The need for standing CSC2 organizations is driven by the Air and Space Expeditionary Force (AEF) environment. In globally responding to threats, AEF combat support resources may need to be allocated from one theater to another to make best use of available resources. Currently, some resources—including theater-based munitions and war reserve materiel, intratheater distribution resources, and physical and operational infrastructures—are primarily confined to individual theaters and are managed by theater-based organizations. For a large number of resources, this arrangement may still prove effective. Nevertheless, the ability to relocate and reallocate these resources to other areas of responsibility (AORs) needs to be streamlined. Other combat support resources are currently managed by units. With the advent of centralized intermediate repair facilities (CIRFs), and to allocate scarce resources, these resources may need to be managed from a global perspective. Other scarce resources that may need to be managed centrally include spare parts, fuel, munitions, aerospace ground equipment (AGE), fuels mobility support equipment, consumables, maintenance, and intertheater distribution resources.

In the remainder of this appendix, we address three "new" standing organizations and their roles in the TO-BE CSC2 operational architecture: the Operational Support Center (OSC), the Commodity Control Points (CCPs), and the Air Force–level Contingency Support Center (CSC).

The Operational Support Center

Integral to the implementation of the CSC2 operational architecture is the OSC. OSCs will provide Air Component Commanders theater-wide daily situational awareness and C2 of air and space; intelligence, surveillance, and reconnaissance; information operations; and mobility, combat, and support forces. The OSC will have the ability to direct deliberate planning and crisis-response actions and to deploy and sustain forces across the spectrum of operations. Within the OSC, the A-4 division will act as the regional hub for monitoring, prioritizing, and allocating theater-level combat support resources. The A-4 will be responsible for providing mission support and base infrastructure support and for establishing movement requirements within the theater. The OSC A-4 will be the theater integrator for commodities managed by Commodity Control Points (discussed in the following section).

To be effective, theater resources must be completely visible to the OSC A-4 and the OSC A-4 must have authority to reconfigure those resources. The A-4 should have the ability to receive commodity-specific information from commodity inventory managers and to perform integrated capability assessments, of both sortie production and base assessments, and to report those capabilities to the combat support personnel supporting air campaign plan/master air attack plan (MAAP)/Air Tasking Order (ATO) production in the Air Operations Center (AOC). In this role, the OSC A-4 will make resource-allocation decisions when there are competing demands for resources within the theater.

In the spares area, the Air Force has made good progress in establishing some of these capabilities in the Regional Supply Squadrons (RSSs). The C2 features of the RSS can be accessed virtually, by computer, by the COMAFFOR A-4 within the OSC. Similarly, in the ammunition area, the theater ammunition control points can provide virtual assessment capabilities to the COMAFFOR A-4. As prescribed in *Command and Control* (USAF, 2001, p. 31), the OSC A-4 could perform these reachback functions. The OSC A-4 could be devoted to incorporating capability assessments of mission, base infrastructure, and movement into operational plans and to supporting the deployed AFFOR A-4 staff during a contingency. These functions would minimize the number of personnel required to deploy forward. It would also alleviate problems associated with an undermanned numbered Air Force staff currently trying to perform the functions listed above as well as their roles under the unified command structure. One example of an OSC already established is the USAFE Theater Air Support Center (UTASC).

The Commodity Control Point

Commodity Control Points should be responsible for making sure that the needed resources are supplied to the major commands (MAJCOMs) and deployed forces, thereby ensuring that critical resources are properly managed and distributed. For example, spares management should be accomplished, along weapons-system lines, by a CCP at the Air Force Materiel Command (AFMC). This standing C2 node at AFMC would manage spares along the continuum of operations, since it would have immediate access to both the data and analytic tools needed to exercise capability assessments and manage distribution of resources to MAJCOMs and theaters. When demands exceed supply, the Spares CCP would take direction from the

Air Force–level CSC—a neutral integrator for arbitrating resource allocations among competing AORs and COMAFFORs.

The Spares CCP would be responsible for monitoring resource inventory levels, locations, and movement. The data would be used to assess contractor and depot capabilities to meet throughput requirements.

The Contingency Support Center

The Air Force Contingency Support Center (CSC), located at the Pentagon, would use operational capability assessments of weapons systems and coordinate with the joint community and theater OSCs to prioritize and allocate resources in accordance with theater and global priorities. These integrated assessments would support allocation decisions when multiple theaters were competing for the same resources and could serve as the Air Force voice to the Joint Staff when arbitration across services is required. In light of the global nature of AEFs and worldwide commitments, other commodities should be considered for management in the same manner.

At both the OSCs and the Air Force CSC, individual resource prioritization will be guided by a common set of rules:

- Given a required operational capability, the OSC will calculate the combat support resources needed to meet the requirement within the AOR.

- When there are multiple ways to achieve the same goals, resources will be assessed and allocated to meet the operational capability requirements that have been prioritized at higher levels (for example, the Joint Chiefs of Staff [JCS] and Air Force CSC).

- Resources will be allocated according to the need for an overall level of operational capability rather than according to the need for an individual commodity.

On the basis of these assessments and allocations, the CCPs will direct purchases, repair operations, and distribution of components and spares. The CCPs will also interface with combatant commanders and the joint community to coordinate intertheater airlift and direct the distribution of resources among theaters. Theater OSCs will advise of infrastructure capabilities, resources needed to implement plans, and the consequences of not improving capabilities. Then the theater joint command can prioritize needs and advise the Joint Staff and others of theater capabilities and issues. Ongoing capability assessments generated by the Air Force CSC and OSCs will be incorporated into a theater's operational planning processes executed by Combat Support Liaisons in the AOC.

CONCLUDING REMARKS

Although these responsibilities can be performed by different organizations in different theaters, the grouping of the tasks, the information required to complete the tasks, and the products resulting from each task should not change from one theater to the next. Predefining the organizations to perform each task will ensure responsibility for tasks, clear lines of communication, and, thus, a smoother transition as the level of operations expands and contracts.

The TO-BE operational architecture was not completed until September 2001, just as OEF began. The architecture had not been completely vetted to senior leadership, so it was not fully implemented during OEF. However, OEF offered another opportunity to examine the processes of the TO-BE architecture. Experiences from OEF have been incorporated into the TO-BE architecture, which has now been vetted and is in the process of being implemented.

CSC2 IN THE CENTRALIZED INTERMEDIATE REPAIR FACILITY TEST

The centralized intermediate repair facility (CIRF) test, which was ongoing during Operation Enduring Freedom (OEF), demonstrated the essential elements of the combat support execution planning and control (CSC2) closed-loop control system. It showed the practicality and benefits of using such a system in contingency operations. Figure B.1 describes the basic features of the closed-loop process used during the CIRF test and described in the CSC2 operational architecture (Appendix A).

The process begins, as shown on the left side of the figure, with the development of an integrated operational and combat support plan. This plan specifies the operational measures of effectiveness (MOEs) to be achieved through combat support activities—for example, F-15 weapons system availability objectives. Performance control parameters based on these operational MOEs are defined for combat support processes to create the desired operational MOE—for example, maintenance repair times, transportation times. The jointly developed plan is then assessed to determine its feasibility according to availabilities of combat support resources. If the plan is infeasible, operational and/or combat support portions of the plan are identified for replanning, as shown by the closed-loop[1] planning portion of the process on the left side of the figure.

---

[1]A *closed-loop process* takes the output and uses it as an input for the next iteration of the process.

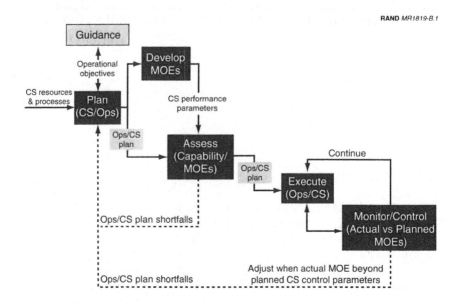

**Figure B.1—The Closed-Loop Process Used to Control Fighter CIRF
Operations in OEF**

Once a feasible plan is established, the jointly developed plan is then executed. In the execution portion of the process, actual performance of a combat support process is compared to the process control parameters that were identified in the planning process, as shown in the lower right of the figure. When a combat support parameter is not within the limits set in the planning process, combat support planners are notified that the process is outside accepted control parameters so that plans can be developed to get the process back within control limits.

The process centers on integrated operational/combat support planning and incorporates activities for continually monitoring and adjusting performance. A key element of planning and execution in the process template is the feedback loop, shown by the output being fed back in as input, which determines how well the system is expected to perform (during planning) or is performing (during execution) and warns of potential system failure. It is this feedback loop that tells the logistics and installations support planners to act when the combat support plan and infrastructure should be reconfigured to meet dynamic operational requirements, both during planning and during execution. Combat support organizations need to be flexible and adaptive so that they can make changes in execution in a timely manner.

In addition to driving changes in the combat support plan, the feedback loop might call for a shift in the operational plan. For the combat support system to provide timely feedback to the operators, it must be tightly coupled with their planning and execution processes and systems and provide options that will result in the same operational effects yet cost less in terms of combat support: Feedback might include notification of missions that cannot be performed because of combat support limitations (Leftwich et al., 2002).

This CSC2 process was implemented during the CIRF test that occurred during OEF. During OEF, F-15 and F-16 units deployed to Operation Northern Watch sites in Turkey, and Operation Southern Watch/OEF sites in Southwest Asia were supported by CIRFs located at Lakenheath, Spangdahlem, and Aviano Air Bases in USAFE. These CIRFs performed repairs on engines (intermediate maintenance), Low Altitude Navigation and Targeting Infrared for Night (LANTIRN), electronic warfare (EW) pods, and F-15 avionics for deployed units. Using this repair concept requires the deployed units to remove and replace unserviceable components from the aircraft at the deployed sites and then to ship the reparable components to the CIRF for repair. The CIRF repairs the reparables and returns serviceable components to the deployed units (see Figure B.2).

As illustrated in Figure B.2, CIRF planners used operational objectives for sortie generation and weapons system availability to establish control parameters for combat support performance, including expected unit component removal rates; transportation times to and from the CIRFs; CIRF repair times; inventory buffer levels—for example, Readiness Spares Package levels; and other parameters. During the CIRF test, actual logistics pipeline performance was tracked against these control parameters.[2]

---

[2]Methods for deriving logistics performance parameters from operational metrics for reparable components have been known for some time. See such publications as Tripp (1983); Tripp and Pyles (1983); and Isaacson and Boren (1993).

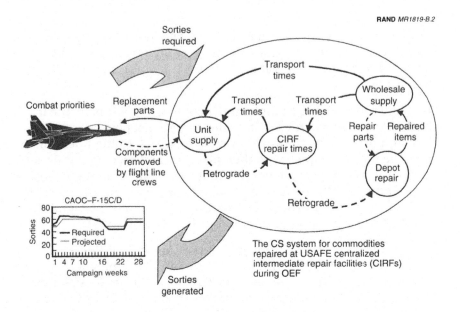

Figure B.2—Combat Support Performance Parameters Were Related to Operational MOEs

The bottom of Figure B.3 shows some of the combat support process control parameters that were monitored. The top half of the figure shows how two parameters associated with Customer Wait Times (CWTs), one from the CIRF to deployed units and the other from depots to the CIRFs, were monitored against transportation performance thresholds, or control limits. The CWT control graphs show the percentiles of total CWT for a number of FOLs for a three-month period in OEF.

The line on each graph illustrates how a control limit could be breached and indicates that the performance of the actual theater distribution system or strategic resupply system was beyond tolerance and, if not corrected, would affect objectives for weapons system availability. This comparison of actual performance against control parameters established from operational goals took place during the OEF CIRF test. Personnel at the USAFE Regional Supply Squadron (RSS) monitored transportation, maintenance, and supply

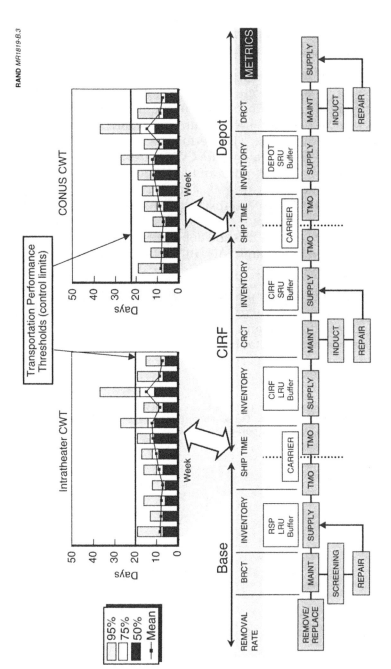

Figure B.3—Actual Process Performance and Resource Levels Were Compared with Planned Values

parameters as shown in this figure against those that were needed to achieve operational weapons system availability objectives.[3]

As shown in this illustration, when the performance of the TDS against these goals was out of tolerance, RSS personnel then indicated, before negative effects could occur, how this performance, if left uncorrected, would affect future operations. As an example, during the OEF CIRF test, TDS performance to Al Jabar Air Base in Kuwait was consistently above the CWT performance criteria of 4 to 6 days for support of EW pods and LANTIRN to this location. The RSS personnel worked with AMC and USTRANSCOM personnel to improve TDS performance to this location; however, CWT could not be improved with resources that USTRANSCOM was willing to allocate for TDS. As a result, RSS and deployed-unit personnel made the decision to deploy EW and LANTIRN repair capability to Al Jabar during the OEF CIRF test.

The use of this closed-loop process during the CIRF test represented a significant improvement in CSC2. These closed-loop concepts, and the associated doctrine and educational programs to fully describe the process, are being established for implementation across a variety of combat support processes Air Force–wide.

----

[3]The RSS personnel were performing a COMAFFOR A-4 function as outlined by the CSC2 operational architecture. These personnel could be considered to be a virtual extension of the UTASC, an Operational Support Center, as described in the CSC2 operational architecture.

A FRAMEWORK FOR ASSESSING SUPPORT CAPABILITIES

This appendix addresses how combat support capability-based assessments can be conducted to provide insights into some very fundamental questions, including (Snyder, 2003):

1. What combat support capabilities exist in the Air and Space Expeditionary Force (AEF) to open bases after steady-state rotational commitments have been met?

2. What combat support capabilities exist in the AEF to augment capabilities at preexisting yet not fully developed bases?

3. What are the limiting constraints on capability, equipment, and personnel?

4. What are the options for mitigating the constraints and what are their costs?

Senior defense planners and senior Air Force officers have laid out the requirement to provide such assessment, and RAND Project AIR FORCE has a project under way that develops these capabilities.

Figure C.1 shows a high-level schematic of methods that have recently been developed to answer these questions (Snyder, 2003). Air Force personnel and equipment resources are inputted into the model, then the current AEF posturing of deployable resources into the AEF buckets is used as a starting point. This process identifies nondeployable resources that are needed to support home-station

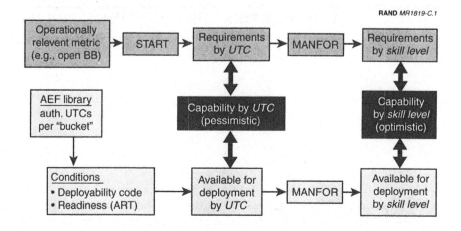

Figure C.1—Model of AEF Capabilities and Options Analysis

(garrison) operations, training, and other readiness requirements. Given the resources identified in the deployable AEF buckets from both a personnel and equipment standpoint, the model then computes the capabilities for opening other forward operating locations (FOLs), given all combat support unit type codes (UTCs) necessary to provide those capabilities.

As shown at the bottom of the figure, this model can be used for evaluating options for satisfying deployment requirements if the desired capabilities are considered inadequate. One of these options would consider substituting contractors to provide home-station or deployment needs. Another might look at deployment concepts more closely aligned to capabilities associated with how forces are presented. In the fuels area, for instance, it takes a relatively large contingent of people, approximately 25, to open a bare base. The activities would include opening fuels mobility support equipment; dragging bladders; hooking up hoses, pumps, and additive stations; building berms; and so forth. Once the fuels layout is developed, fewer than 25 people are needed to operate the fueling operations.

Current UTCs are not built along these capabilities lines. A fuels deployment UTC contains both the build components and the operate components. The reality is that once a fuels UTC deploys, it stays deployed. Changing the UTCs to be capability-based provides the opportunity to extract capabilities that are no longer needed and to reuse them, thereby freeing up some resources that are not required for the entire deployment cycle.

This capability-based assessment method also determines the equipment UTCs that are needed to develop an FOL to support a wide variety of force packages, threat conditions, and available infrastructures at the deployed site. Figure C.2 illustrates how the method would determine the UTCs that are needed to develop a site to bed down and operate an 18-PAA (Primary Aircraft Authorized) fighter squadron at a bare base, with a given threat and other factors. The method also identifies the limiting UTC equipment constraint, the most-constrained UTC equipment and, therefore, indicates how many 18-PAA fighter squadrons can be opened, given the binding constraint.

The method also determines manpower constraints. Using the same example of assessing the number of 18-PAA-fighter bare bases that can be opened, Figure C.3 shows that the manpower constraint can be bounded, on the one hand, on the optimistic side by considering

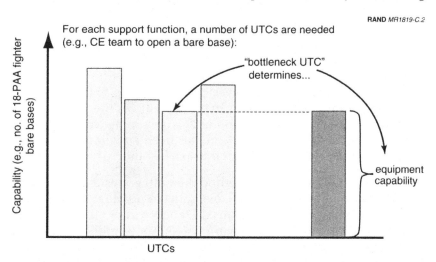

NOTE: Bottleneck is "limiting equipment constraint" cited in the text.

Figure C.2—Notional Equipment Capabilities and Constraints Model

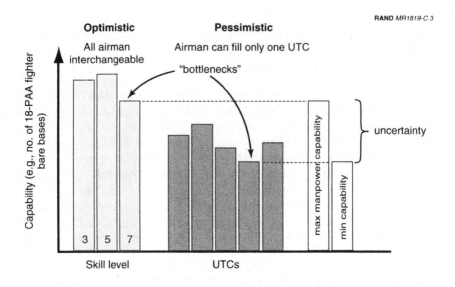

NOTE: Bottleneck is "limiting factors" cited in the text.

Figure C.3—Notional Manpower Capabilities and Constraints Model

the available skill levels associated with a particular combat support area—for example, fuels support. This optimistic bound assumes that all airmen can satisfy requirements specified in deployment UTCs. In this case, the seven levels postured in the AEF library constrain the number of bases that can be opened. On the other hand, a pessimistic bound can be determined by assuming that an airman can fill only the UTC that he or she has been assigned. These two views bound the number of bases that can be built.

The capabilities-assessment method then integrates the equipment constraints and the manpower constraints to provide a view of how much capability exists within an AEF bucket to open and operate FOLs of various categories. Figure C.4 shows (notionally) that the manpower constraint is the binding constraint in building FOLs. The *residual capability* is the number of FOLs that can be built after the steady-state requirements are satisfied. This method, therefore, can assess and describe the combat support capabilities that lie within AEF buckets that are constructed using various rules.

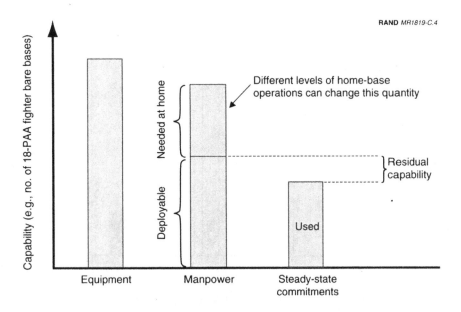

Figure C.4—Notional Model of the Integration of Equipment and Manpower Capabilities of AEF Buckets

We use fuels manpower as an example to demonstrate how the combat support capabilities assessment method works. Figure C.5 shows the AEF capability to open and operate 18-PAA bare bases. The 3-skill level in the fuels area provides the upper bound on the number of bare bases that can be provided using current rules for determining deployable Air Force specialty codes, the DXS or DWS code in the AEF library database. The people in the JFABA UTC determine the lower bound. The far right-hand bar indicates the capability that is used up in satisfying the pre-9/11 steady-state rotations. After the rotational commitments have been satisfied, the residual capability to open 18-PAA bare bases within an AEF bucket to meet contingency requirements lies between one and two bases.

The method begins to provide the framework to answer the capability-based questions that the senior defense planners and senior Air Force leaders have asked.

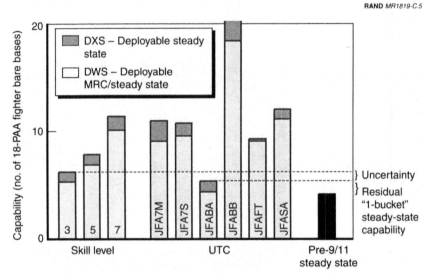

RAND *MR1819-C.5*

NOTE: Abbreviations for UTCs are Air Force Specialty Codes.

Figure C.5—Fuels Capability During the Steady State Is Limited

BIBLIOGRAPHY

Amouzegar, Mahyar A., Lionel A. Galway, and Amanda Geller, *Supporting Expeditionary Aerospace Forces: Alternatives for Jet Engine Intermediate Maintenance*, Santa Monica, Calif.: RAND Corporation, MR-1431-AF, 2002.

Barthold, Col Bruce R., "Major Issues from the AFCESA/AFIT Sponsored Operation Enduring Freedom RED HORSE and PRIME BEEF Lessons Learned Conference, 13–15 Nov 02," presented at conference, Dayton, Ohio, December 2002.

Boyd, John R., "A Discourse on Winning and Losing," Maxwell AFB, Ala.: Air University Library, Document No. M-U43947, unpublished collection of briefing slides, August 1987.

DynCorp, "Operation Enduring Freedom Oct 01–Mar 02," briefing to DynCorp staff, Shaw AFB, S.C., n.d.

Feinberg, Amatzia, H. L. Shulman, L. W. Miller, and Robert S. Tripp, *Supporting Expeditionary Aerospace Forces: Expanded Analysis of LANTIRN Options*, Santa Monica, Calif.: RAND Corporation, MR-1225-AF, 2001.

Galway, Lionel, Mahyar A. Amouzegar, R. J. Hillestad, and Don Snyder, *Reconfiguring Footprint to Speed Expeditionary Aerospace Forces Deployment*, Santa Monica, Calif.: RAND Corporation, MR-1625-AF, 2002.

Galway, Lionel, Robert S. Tripp, Timothy L. Ramey, and John G. Drew, *Supporting Expeditionary Aerospace Forces: New Agile*

Combat Support Postures, Santa Monica, Calif.:   RAND Corporation, MR-1075-AF, 2000.

Geller, Amanda, David George, Robert S. Tripp, Mahyar A. Amouzegar, and Charles Robert Roll, "Supporting Air and Space Expeditionary Forces: Analysis of Maintenance Forward Support Location Operations," Santa Monica, Calif.: RAND Corporation, unpublished RAND research, 2003.

Halliday, John, and Nancy Y. Moore, *Materiel Distribution: Improving Support to Army Operations in Peace and War*, Santa Monica, Calif.: RAND Corporation, IP-128, 1994.

Headquarters U.S. Air Force (HQ USAF), External IL Slides, daily update to CSAF, September 19 and 20, 2001a.   Posted at www.afxo.pentagon.smil.mil.

HQ USAF, *USAFE/SA Air War Over Serbia Report*, April 2000.

HQ USAF, Operations Group briefing to CSAF, October 30, 2001b. Posted daily at www.afxo.pentagon.smil.mil.

Hillestad, Richard, Robert Kerchner, Hyman Shulman, and Louis Miller, "The Closed Loop System for Planning and Executing Weapon System Support," briefing to RAND staff, January 13, 2003.

Hornburg, Gen Hal M., "ACC Issues," briefing to RAND, Langley AFB, Va., July 2002.

Isaacson, Karen E., and Patricia M. Boren, *Dyna-METRIC Version 6: An Advanced Capability Assessment Model*, Santa Monica, Calif.: RAND Corporation, R-4214-AF, 1993.

Joint Chiefs of Staff, Office of the Chairman, *Joint Tactics, Techniques, and Procedures for Movement Control*, Washington, D.C., Joint Publication 4-01.3, 1996.

Jumper, Gen John P., Air Force Chief of Staff, "Chief's Sight Picture," Washington, D.C., May 3, 2002.

Jumper, Gen John P., "The Limitations of Doctrine," briefing given to Air Force Doctrine Center, Maxwell AFB, Ala., n.d.

Lambeth, Benjamin S., *NATO's Air War for Kosovo: A Strategic and Operational Assessment*, Santa Monica, Calif.: RAND Corporation, MR-1365-AF, 2001.

Leftwich, James A., Robert S. Tripp, Amanda Geller, Patrick H. Mills, Tom LaTourrette, Charles Robert Roll, Cauley Von Hoffman, and David Johansen, *Supporting Expeditionary Aerospace Forces: An Operational Architecture for Combat Support Execution Planning and Control*, Santa Monica, Calif.: RAND Corporation, MR-1536-AF, 2002.

Martin, Gen Gregory, Commander, U.S. Air Forces, Europe (USAFE), "USAFE Role in Operation Enduring Freedom," white paper, January 15, 2002.

Nardulli, Bruce R., Walt L. Perry, Bruce Pirnie, John Gordon, and John G. McGinn, *Disjointed War, Military Operations in Kosovo, 1999*, Santa Monica, Calif.: RAND Corporation, MR-1406-A, 2002.

Peltz, Eric, H. L. Shulman, Robert S. Tripp, Timothy L. Ramey, Randy King, and John G. Drew, *Supporting Expeditionary Aerospace Forces: An Analysis of F-15 Avionics Options*, Santa Monica, Calif.: RAND Corporation, MR-1174-AF, 2000.

Polich, J. Michael, Bruce R. Orvis, and W. Michael Hix, *Small Deployments, Big Problems*, Santa Monica, Calif.: RAND Corporation, IP-197-A, 2000.

Przybyslawski, Brig Gen Tony, "AEF Weekly Metrics," AEF Center update to CSAF, November 12, 2002. Posted at www.aef.Langley.af.smil.mil.

Pyles, Raymond, and Robert S. Tripp, *Measuring and Managing Readiness: The Concept and Design of the Combat Support Capability Management System*, Santa Monica, Calif.: RAND Corporation, N-1840-AF, 1982.

Rosenthal, Maj Robert, "Air Force Operations Group Noble Eagle/Enduring Freedom Operations Update," briefing, January 15, 2002.

Snyder, Donald, "Analysis of Options for Presenting Forces and Managing Garrison Capabilities Within the AEF Construct," RAND briefing to the AEF Center, Langley AFB, Va., January 13, 2003.

Sortor, Ronald E., and J. Michael Polich, *Deployments and Army Personnel Tempo*, Santa Monica, Calif.: RAND Corporation, MR-1417-A, 2001.

Tripp, Robert S., "Measuring and Managing Readiness: The Concept and Design of a Wartime Spares Push System," *Logistics Spectrum*, Vol. 17, No. 2, Summer 1983.

Tripp, Robert S., Lionel A. Galway, Paul S. Killingsworth, Eric Peltz, Timothy L. Ramey, and John G. Drew, *Supporting Expeditionary Aerospace Forces: An Integrated Strategic Agile Combat Support Planning Framework*, Santa Monica, Calif.: RAND Corporation, MR-1056-AF, 1999.

Tripp, Robert S., Lionel A. Galway, Timothy L. Ramey, Mahyar A. Amouzegar, and Eric Peltz, *Supporting Expeditionary Aerospace Forces: A Concept for Evolving to the Agile Combat Support/ Mobility System of the Future*, Santa Monica, Calif.: RAND Corporation, MR-1179-AF, 2000.

Tripp, Robert S., and Raymond Pyles, "Measuring and Managing Readiness: An Old Problem—A New Approach," *Air Force Journal of Logistics*, Spring 1983.

U.S. Air Force, *Combat Support*, Washington, D.C., Air Force Doctrine Document 2-4, October 2000.

U.S. Air Force, *Command and Control*, Washington, D.C., Air Force Doctrine Document 2-8 (Draft), February 16, 2001.

U.S. Air Force, *Organization and Employment of Aerospace Power*, Washington, D.C., Air Force Doctrine Document 2, September 28, 1998.

U.S. Air Force, 437th Airlift Wing, "Operation Enduring Freedom, 437th OEF Contributions," fact sheet, Charleston, S.C., May 22, 2002.

U.S. Air Force, Air Force Operations Group, "CAT Update," briefing, January 14, 2002.

U.S. Air Force, Air Mobility Command (AMC), "Mobility Air Forces Tanker Airlift Control Center," briefing, Scott AFB, Ill., March 13, 2002.

U.S. Transportation Command (USTRANSCOM), USTRANSCOM Update Brief, January 14, 2002. Posted at www.TRANSCOM. smil.mil.

Wolff, Robert D., *Contingency Contracting in Support of Air Expeditionary Forces*, report to RAND under Contract No. F49642-96-C-001, December 29, 2000.

Made in the USA
Las Vegas, NV
29 April 2022